SITE DESIGN & ENGINEERING

A SITE DEVELOPMENT, CONSTRUCTION PLANS PREPARATION, AND PERMITTING GUIDE

ALI SHASTI, P.E.

To request permission, contact the author, M. Ali Shasti-Nazem
Professional Engineering Services LLC
P. O. Box 58573
Renton, WA 98058
AShasti13@Gmail.com
www.My-PE.com

Cover Design and Formatting by
Word-2-Kindle
Edited by Joanne Lane,
First Editing

ISBN 979-8-9870937-6-4
First Edition

Printed in the United States of America
IngramSpark

Dedication

I dedicate this book to my parents, who worked very hard throughout their lives to make ends meet, and to all decent and hardworking Americans and Iranians, including my family members in both countries. Many, like my parents, work honestly and diligently and try every day to meet the numerous challenges of their daily lives.

I also dedicate this book to Iran, a country I grew up in and left when I was twenty-six. I was educated in Iran and learned to be hardworking, disciplined, and respectful. America further advanced my education and taught me to be an ethical professional, fair, resourceful, and confident to author this book. They are two countries that I love, despite all their shortcomings, pitfalls, and imperfections.

Acknowledgements

I want to thank Clem Trautwein, the senior designer working for Access Engineering and Consulting, Inc. for over twelve years, and for designing and drafting many of the drawings cited in the book. I also thank Eileen Shirley, the TampaBayCADD, who reworked and manipulated many of these drawings to format them into smaller formats on pages. In addition, thanks to Peter Schwarz, Realtor with This Sold House Realty, Inc., Scott Smith, P.E. with the King County Department of Permitting and Environmental Services, and Mike Lewis, PLS, State of Florida Professional Land Surveyor, who all participated in the review of a specific portion of this book.

Special thanks to my associate of three years Grace Amundsen Barnkow, P.E., current Pasco County Engineer, for her support, review, and providing valuable comments.

Many thanks to Joanne Lane of FirstEditing.com, who made handy suggestions and comments on bettering this book. I also thank all those who I was lucky enough to cross their path and who provided an opportunity for me to learn from them.

Finally, I am grateful to have such a wonderful wife, children, and cousin. They always supported and encouraged me in my efforts to succeed in my career, including authoring this book.

Disclaimer

The information, tables, drawings, and other materials provided in this book are general informational purposes and are based on the engineering knowledge and experience of the author. All information is provided in good faith to guide readers to experience successful design, construction plans, documentation, and permitting experiences. However, the author makes no representation or warranty of any kind, express or implied, regarding the accuracy, adequacy, validity, reliability, availability, or completeness of any information in this book.

YOUR USE OF THIS BOOK AND YOUR RELIANCE ON ANY INFORMATION PROVIDED IN IT IS SOLELY AT YOUR OWN RISK. UNDER NO CIRCUMSTANCES SHALL WE HAVE ANY LIABILITY TO YOU FOR ANY LOSS OR DAMAGE OF ANY KIND INCURRED AS A RESULT OF THE USE OF THIS BOOK OR RELIANCE ON ANY INFORMATION PROVIDED IN IT.

TABLE OF CONTENTS

INTRODUCTION

There are more than three thousand counties and close to twenty thousand cities in the United States of America. Ten cities have populations above one million. About three hundred cities are considered medium cities with populations of 100,000 or more. Approximately fifteen thousand cities, towns, and villages have people below 5,000.

All these municipalities follow the rules and regulations established by their state laws and the Federal government. They also develop other rules and regulations, including engineering and design standards, that could be different or sometimes more stringent than those of other agencies.

Although many rules and regulations have similarities, there are no uniform standards among the agencies mentioned above. Each of the organizations mentioned above has its own rules and regulations that must be adhered to for permit-related activities. Therefore, it is almost impossible to have a document to capture all requirements that pertain to every conceivable situation of a particular agency having jurisdiction (AHJ).

The author's primary purpose is to provide much-needed residential and commercial site development guidance to property owners, developers, architects, engineers, students, and contractors to develop a sound set of approvable and constructive construction plans. It also intends to explain the permitting process, fast-tracking permit approval, and development of sound engineering projects. This book attempts to answer hundreds of questions the developers and private engineering consultants asked during interactions between the author and the public during his four decades of engineering and consulting. The questions mentioned above surfaced during pre-application meetings through the review and approval process to complete the project and submit the as-built drawing by the engineer of the record (EOR).

Federal and state governments and municipalities have numerous rules and regulations about site development. Some cities have policies and procedures that differ from federal and state governments by adopting more stringent criteria through local ordinances. This book's information may help readers understand various agencies' rules and requirements, communicate more effectively with all parties involved, and negotiate to their advantage.

Experienced engineers, design practitioners, and developers may also use this book as a refresher and keep track of their progress by following some of the procedures outlined here.

How to use this book

The **first chapter** evaluates a suitable site for the type of development a potential property owner may seek. It explains how to select a property with the least number of problems and risks. And how to be mindful of potential issues to avoid ensuring profitability. Acquiring a valuable property sets the stage for the rest of the chapters in the book.

The **second** and **third chapters** focus on gathering the survey and geotechnical information for the site. Although surveying differs from geotechnical engineering, they are two of the essential elements of any successful design and construction project. They also look at what data and information to seek when ordering a survey and a geotechnical report.

Chapter 4 provides some design information and processes after the first three chapters. There are many ways of designing a specific project. However, there probably aren't too many ways to optimize a design. This chapter aims to provide a basic understanding of design elements and processes, hoping to achieve the most efficient method in any site condition.

Chapter 5 reviews different activities and their definitions and requirements for construction and other activities in the right-of-way. Understanding how to approach the AHJ and obtain the necessary permits for encroaching right-of-way for construction and non-construction activities is essential.

Chapter 6 provides information regarding single-family residential and accessory dwelling units (mother-in-law) and how to prepare construction plans to start the permitting process. It looks at how to prepare site engineering construction plans and what to submit to the AHJ. This type of permit is probably the most common, since there is always a need for additional homes. As of March 2022, a shortage of four million homes exists in the United States of America. This trend will probably continue for years and even decades, for various reasons outside the scope of this book.

Chapter 7 discusses the design and construction plans prepared for a small sixteen-lot residential subdivision. Project 2, featured in this chapter, shows how a project starts from vacant land and develops to the survey map and a final completed construction plan—the only example in this book that provides comprehensive design and construction plans for the entire project. For readers interested in the design, construction plans preparation, and permitting, I recommend reading or studying this chapter before proceeding to chapters eight and nine.

Chapter 8 is the same as chapter seven, except it is an example of a larger nineteen-lot residential subdivision. Plus, each lot is much larger than the lots in Project 2. Therefore, it has different challenges reviewed and discussed in this chapter. Readers are encouraged to read or study chapter seven before venturing into this and the following chapter.

Chapter 9 presents commercial project site improvements for a three-story chain hotel and a potential dine-in restaurant. Even though there are similarities between this project's stormwater management system with the previous projects, settled differences are essential to pay attention to in commercial sites—differences in the parking lot design and other facilities such as commercial swimming pools and spas. This chapter discusses all those elements specific to a hotel site design. The author recommends the reader read or study chapter seven before venturing into this chapter.

Chapter 10 reviews and discusses the permitting agencies' processes and requirements. Since numerous permitting agencies have many provisions that could significantly differ from others, covering all these requirements is almost impractical and impossible. However, covering many general requirements and processes provides some ideas about the needs and procedures, which helps the permittees have realistic expectations.

CHAPTER 1: SITE EVALUATION

1.1. Introduction

This chapter discusses the importance of selecting a site location. Choosing a site location is one of a potential developer's first decisions. Accessibility, topography, soil conditions, and shape factor are all the elements to be considered. Other aspects to consider include setbacks and frontage improvement requirements. The above-stated features can allow developers to select a better site with the best return on investment and an acceptable amount of risk.

Understanding these factors can also be beneficial for an already owned or purchased property. It can serve to identify limitations to factor in deciding how to develop to maximize the returns.

This chapter intends to present a description and understanding of selecting a site.

1.2. Zoning and Land Use

Zoning is a tool that government agencies use to regulate development and help implement the goals of their comprehensive plan. Zoning refers to the city's or county's division into various zoning districts according to an official zoning map where the rules for each zoning district are outlined in the Land Development Code (the "zoning code"). Zoning determines the specific land use such as residential, office, commercial, and industrial. In addition, it establishes the smallest lot size and dimensional standards, such as the minimum building setbacks and the maximum building heights. The zoning code also includes regulations governing off-street parking, landscaping, signs, fences, accessory structures, and the creation of subdivisions.

Zoning and land use designations are the first items checked in any land transaction to ensure that the property is zoned for the tenant's intended purpose. As mentioned earlier, the local agencies have zoning, and land use maps to share the exact allowable zoning and land use with their customers. There are many books and articles written about how the local government agencies come up with those maps outside the scope of this book.

Applicants may contact the local agencies planning or zoning department to get their property zoning and land use designations. The best approach is to contact the local government agencies and the zoning department by providing the property parcel number and physical address obtained from the real estate agent, helping to acquire the property or the municipality website. Then, call the local planning or zoning of the AHJ and ask for that information.

The rezoning process is time-consuming and expensive, assuming that the result is positive. For example, if a property is zoned residential, and the plan is to use it for industrial use, it may be wiser to find another property zoned industrial. Even if the goal is to use it for residential purposes, it should be checked to ensure that the land use is allowed for the type of use. Single family residential, apartment complexes, and town-homes all are zoned residential with different land-use types. Thus, by contacting the appropriate department of the local government agencies, one makes sure that the zoning and land use designations are aligned with the proposed intended use and the scope of service for the project.

1.3. Change of Use

Change of use is when a current owner or the new buyer of an existing building proposes to change the use. For example, if the building is currently being used for general office and the new tenant is a dentist and offers using the building as a medical-dental office. On the surface, it may seem they both are office use, and there should not be any permitting required. On a deeper level, there are two concerns from the AHJ's perspective:

1. Does the inside environment meet the new building code and be safe for the new operation? This area is outside of the scope of this book. I only emphasize the applicant contact with the local AHJ to see what building-related permits, such as electrical, plumbing, etc., need to be obtained based on the complexity of the new operation.
2. A medical-dental office generates more traffic than a general office. Therefore, most permitting agencies require additional transportation impact fees for this use change. The transportation impact fees are usually based on one or a combination of the following methods as appropriate:

    a. the AHJ fee schedule,
    b. *Institute of Transportation Engineers (ITE) Manual,*
    c. Traffic Impact Analysis (TIA) performed by a qualified transportation engineering firm.

Of course, any methods mentioned above need to be verified and approved by the AHJ over the project site.

1.4. Location and Access

Those familiar with real estate have already heard of the phrase location, location, location. Educated and experienced developers sometimes pursue buying properties with many challenges. The purpose is to negotiate a much lower price. Of course, the best location also often comes at the highest cost.

Such properties include, but are not limited to, sites with flooding conditions, steep slopes, wetlands, creeks, or other sensitive areas. In addition, properties with no direct access or an easement connection to roads and sites face many extraordinary challenges. Such properties present significant extra design hurdles. They may also need more complex environmental and other governmental agencies permitting requirements. Yet, their potential return on investment could be higher when adequately chosen and developed. The developers realize that the cost savings of purchasing such property may outweigh all other challenges. Of course, upon performing a feasibility study and evaluating the pros and cons.

Assuming the real estate market has conditions to allow for enough lead time, an experienced developer will invest in commissioning a feasibility study or as part of the purchasing process. A civil engineer or a design firm with a good reputation would be an excellent choice for conducting research. Paying attention to the elements in this section may help the buyer reach the rate of returns they expect. A feasibility study should cover a lot of these elements. Suppose the timeline for purchasing doesn't allow for a feasibility study; one may research on their own on these elements. Studying the listing, conducting a site visit, or evaluating the site on Google Earth will still help provide a better picture of the relative value of the investment.

Regardless, check accessibility such as driveway locations, availability, sizing restrictions, and options for building a private access road. Before securing the place, confirm accessibility to the site with the municipalities, counties, or other AHJ.

It is also important to remember that access for an oversized vehicle such as semi-trailers is much different from regular small to medium-sized passenger cars. Permitting agencies may require wider driveways, extended throat length, and larger return radii for the semi to get to the site without creating traffic back up and other traffic-related problems and unsafe conditions.

What if access does not exist? What is the feasibility of acquiring an access easement from a neighboring property owner? That process can be time-consuming, expensive, and involve lengthy negotiations. Thus, it should be factored into the determination for the location. Access may also pose other challenges. Steep grades that need earthwork or retaining walls to remedy the site conditions. Other options would be establishing new access off another side of the property from a different road.

Some arterial roads may have limits on the number or spacing of driveways. This AHJ rule might make some site configurations difficult or sometimes impossible.

1.5. Site Considerations

There are many ways that developers and contractors approach securing and buying sites. Depending on the experience, background, and objectives of the purchaser, these approaches can be classified into two main tactics to do this task.

1)   design to fit a site or

2)   buy a property, and then design accordingly.

The best time for selecting a site occurs after the scope of services and the design program has been established, and a realistic estimate has been made to meet the owner's space needs. Schematic studies then can show how many buildings be required; how much land will be needed around each building. Areas such as parking, dumpster location, landscape islands and retention ponds, and characteristics desired of the site and surrounding property. Estimates also can be made of utilities needed. The data derived from these studies are basic considerations in selecting a site (Merritt, 1990).

If site selection is approached, the owner can leverage the project budget for the local marketplace conditions. A site might be best developed as a residential subdivision. Still, if the local housing market is saturated, a developer might be wiser instead to choose a parcel better suited to commercial development or at least revamping the project schedule.

One of the first decisions that a developer must make is where to develop. A good place to start is to gather as much information as possible about the site. The following sections will give you some ideas about where to start.

There are some valuable sources of information to answer these questions. They include, but are not limited to, your real estate agent, the local authority having jurisdiction (city, county, or even state). Other valuable sources are the GIS maps from the Federal Emergency Management Agency (FEMA) or the United States Geological Survey (USGS).

If it is possible, walk the site. Get a feel for the surface water drainage. Does anything look like wetlands? Do you have steep slopes or abrupt grade changes? Is water collecting somewhere or coming onto your site from neighboring properties? Is there any sanitary sewer manhole and/or fire hydrant around the place? A site walk can give the buyer ideas for further research before purchasing. A site walk could be accomplished even on an accelerated purchasing schedule in a competitive real estate market.

1.6. Topography

A wooded site with steep slopes is more challenging to develop than a flat land with few trees. Many agencies have restrictions on cutting significant trees. They may also impose strict substitution for removed trees. These requirements may make it difficult to manage more trees. The only other option would be reducing the building envelope.

Steep slopes could be accommodated with retaining walls or rockeries that increase the construction cost. One option would be to reduce building size to accommodate steep slopes to avoid retaining walls or rockeries.

A depressed site could collect stormwater runoff from the neighboring properties. This condition may create challenges to meet the local, state, and federal government requirements. Particularly with those agencies with more strict water quality and quantity requirements. Agencies with more restrictions require a zero discharge to compensate for these conditions, which means extra expenses to design an approvable stormwater system. Zero-discharge means no discharge offsite. To accomplish a zero-discharge system, retention and not detention pond must be designed and constructed to store the project's stormwater runoff. In economic terms, more cost to the project.

1.7. Soil Types and Information

The USGS and local jurisdictional GIS maps can provide general soil information designed for many different users. Users such as farmers, ranchers, foresters, community planners and officials, builders, developers, home buyers, and civil engineers. They can be a good starting point before securing a site.

Significant differences in soil properties can occur within short distances. Some soils are seasonally wet or subject to flooding. Some are too unstable to be used as a foundation for buildings or roads. Clayey or wet soils are poorly suited to use as an on-site sewage facility (septic tank absorption field). A high-water table makes a land poorly suited to the basement or underground installation.

The following soil properties that affect land use are described in the United States Department of Agriculture, Soil Conservation Service (The Soil Survey). Building site development, recreational development, sanitary facilities, construction materials, water management, engineering index properties, physical and chemical properties of soil, soil and water features, and classification of the soils.

The soil survey also contains information regarding the general nature of the county. This information includes history and development, climate, and soil classification. They are listed under engineering index properties, farming, mining, water resources, and transformation within the county based on unified and American Association of State Highway and Transportation Officials (AASHTO). For more detailed information regarding soil properties and their uses, see the *Geotechnical Engineering Part of Standard Handbook for Civil Engineers* by Frederick S. Merritt and William S. Gardner.

1.8. Geometry and Dimensions Ratio

Most people who invest in property, not civil engineers or architects, do not pay much attention to this factor. When a designer lays out future improvements on a survey plan after the developer commits to the property, it usually becomes known. For example, there is much more potential for wasted space at the corners when considering a triangle lot than for a similar area rectangular lot. That is perhaps one main reason that there are fewer triangle buildings. However, a long and narrow rectangle site may restrict the designer when providing an efficient parking design for a rectangle lot. A slim and long lot may be feasible and justifiable for specific uses such as mini warehouses. It is assuming a much lower purchase price for the property.

Regardless, it is highly recommended that a preliminary plan be prepared to avoid any potential geometry issues. It should be based on an existing site survey and incorporate the future scope of work.

1.9. Allowable Floor to Area Ratio (FAR)

The floor to area ratio (FAR) is a standard requirement per the zoning code that must be maintained. This ratio represents how much of a lot can be used for a building, expressed as a decimal number corresponding to the total allowable building area divided by the total lot area. The FAR can be easily obtained by contacting the local government agency or municipality. On occasions, it is as easy as getting on the internet and finding the information on the jurisdiction's website in many areas.

In some exceptional cases, the JHA may grant a variance or waiver from the requirement. However, this option should be explored as a last resort. Applying and securing a variance is time-consuming, risky, and costly. In the end, a variance or waiver may or may not be granted.

Therefore, if the plan is to build 100,000 SF of the office building and the local FAR is 0.5, then at least a site area of 200,000 SF must be available for the proposed facility.

Of course, most often, the required number of parking spaces and other factors such as setbacks and stormwater detention, landscape buffers, and internal landscape islands will be the controlling factors determining the size of the site. It is good to perform this simple computation to check and keep the FAR in mind when choosing allowable use.

1.10. Density and Units Per Acre for Residential Sites

Density applies more to residential sites and involves computing the number of units per acre allowed within a specific zoning district. For example, R-1 usually means one unit of single-family residential on one acre of land. R-4 means four units are allowed on an acre of land.

Some agencies define one-acre land as a percentage of dry or buildable land. Suppose a two-acre parcel has 1.5 acres of wetland. In that case, it is imperative to determine how many units can be built on the half-acre of the remaining dry land instead of assuming that two single-family residential homes are allowed on the two-acre property for the R-1 zone.

1.11. Frontage

Frontage is an essential factor in selecting a site. Frontage refers to the property lines fronting directly on a street or streets. Frontage has several effects on the potential uses and the cost of the investment the site requires.

A site in a prime location with frontage on a major street is likely to be desirable for a wide variety of developments, such as hotels, casinos, restaurants, or retail space that is visible and can get more natural traffic. Frontage on a busy road may be undesirable for residential development. Still, frontage on a side road may save a developer the cost of developing a private road or easement to access the lots.

On the other hand, the price of a property may directly relate to the size of its frontage. Prime frontage carries with it an absolute premium. Therefore, one may want to consider frontage when necessary for the intended site development. Office and apartment complexes, for example, do not need to have long exposures since all they need to provide is adequate space for the driveway(s) accessing the site, a location for a sign, and minimal landscaping, lighting, and similar consideration. On the other hand, the frontage is necessary for an auto dealership site, since viewing not just a sign and the access but seeing the cars from the street is essential.

1.12. Utilities

Another critical factor to consider before securing a site is utilities.

A walk on and around the site reveals many of the essential utilities near the property. Look for stormwater inlets for stormwater connection discharging from the site. Fire hydrant and water valves to provide potable and fire water services to the property. Sanitary sewer manholes nearby the site to handle wastewater discharge generated by the proposed site. Remember that extension of any utilities offsite to the desired utility can substantially add to the development cost. In addition, it developed complexity to the permitting process.

Other utilities such as electric power, gas, cables are also an essential part of any development. Power poles carrying electric lines and other cable lines and gas markers can answer these concerns.

Regardless of a site visit, contacting municipal agencies to get hold of a geographic information system (GIS) may provide at least much of the preliminary information. These days, most progressive agencies own and maintain an excellent GIS map with their utility data.

1.13. Land Cost

Land cost is often a significant factor in deciding which lot to purchase. Land cost receives a lot of attention from developers because it is a substantial commitment of funds. Most of the other factors discussed in this chapter are associated with whether a lot is suitable for building. Those factors must always be weighed against the cost to determine if the investment is sound.

However, good consultants and developers must be aware that the purchase price of a lot is not the only initial outlay of funds. Do not overextend yourself financially by neglecting to account for things like:

- Broker and legal fees,
- Registration fees,
- Title insurance premiums,
- Zoning or subdivision regulations fees,
- Review and permitting fees,
- Provision of access to the site (driveway connection, access management, and easements, sidewalks, right-of-way (ROW) use permits),

- Utility connection fees and the availability of utilities to the site (know where the site would connect to power, sewer, water, and gas to make sure to account for any required service extensions),
- Easement agreement fees and/or cost (considerations such as vacating easements that may interfere with developing the lot to its fullest potential).

The first of the three items mentioned above can easily be obtained by the real estate company working on the property. The rest are more complicated, and most can be obtained by contacting the AHJ and working with the civil engineer associated with the project.

1.14. Feasibility Study or Due Diligence

Often developers and property buyers hire consulting and other professional firms to perform a feasibility study for the property they plan to purchase. The study aims to evaluate all or some of the elements mentioned above and recommends a course of action. Essentially, this study should determine the pro and cons of investing in the subject property and conclude whether purchasing the property due to its profitability is a good investment or not.

For example, one of the AEC clients asked us to perform a feasibility study to purchase a property in a high-end use area to build a three-story chain hotel. After spending some time and gathering various technical information, including geotechnical reports, the AEC staff recommended that the proposed building needs to be constructed on a piling foundation for an approximate cost of $300,000 before continuing with the typical spread foundation. The buyer reevaluated their budget and decided to skip buying the subject property and moved on to another property. Although the potential buyer spent a substantial amount of time and a few thousand dollars for the feasibility study, they determined to move on and accept a few thousand dollars loss instead of risking a few hundred thousand dollars potential loss.

CHAPTER 2: LAND SURVEYING

2.1. Introduction

The importance of a site survey is often underestimated on many projects, particularly on smaller ones. Occasionally, inexperienced owners, developers, and contractors try to save time and money by cutting corners when ordering a site survey. However, the quality of a site survey is related to saving money later through reduced redesign costs and field changes during construction. This chapter discusses the importance of a site survey and evaluates the existing site conditions before starting any design work. There are many options for what to include in a site survey and what level of depth to request. Here in this chapter, we will try to cover the most common ones.

What is Land Surveying?

Land surveying is a scientific technique of determining property dimensions and contours of the earth's surface by measuring the distances, directions, and elevations. Land surveyors use these measurements and physics, mathematics, engineering, and law to establish land boundaries and maps.

All design work performed by civil engineers and architects should be based on actual site surveys, not GIS or sketches. It is paramount to have a complete site survey before starting the design work. Therefore, the EOR, most often a civil engineer, is perhaps the most qualified person to develop the scope of work. They should keep in mind the characteristics of the proposed site design, applicable agencies' requirements, and other site engineering-related items. They pay careful attention to site-specific elements like driveway and retaining wall location options, stormwater design options, zoning requirements, and similar design features.

Construction plans should show property boundaries, ROW lines, easements, topography with contours. A professional land surveyor registered in the state the work is being performed must prepare them. Most municipalities require survey maps as part of the applicant construction plans submitted for any work completed with their buildings, street use, or other permits. All surveys are needed to meet specific minimum standards.

It is recommended that the civil engineer or the surveyor also contact the local governmental agencies (city or county) to obtain the latest Geographic Information System (GIS) maps for the project area. City, county, utility districts, and other agencies maintain GIS maps of their utility facilities, improvements, and other features that can be useful to the surveyor and often reduce the time and cost of performing a site survey without reducing the quality. Upon obtaining the above-stated information, a civil engineer would be able to apply their best knowledge of the site and determine what kinds of surveying to order. In addition, other qualified professionals may develop the same work scope for a registered surveyor to perform the work based on their experiences, regardless of their educational background.

This chapter intends to present a description and understanding of the types of surveys and the minimum requirements for a site survey. All work consisting of the practice of land surveying shall be performed by or under the direction of a surveyor licensed to practice in the state that the work is to be done.

2.2. Boundary Survey

boundary survey should, at least, include property corners, dimensions, and orientation of each property line. It may also provide wetland boundaries and other water bodies located within the site. In that case, typically, each corner is flagged, and the northing/easting is surveyed, inspected, and approved by the environmental agencies having jurisdiction over wetlands or other qualified professionals. It is crucial to remember that it is much easier, less time-consuming, and perhaps less expensive to survey a wetland of regular shapes such as square, rectangle, or circle, than an irregular swamp. Therefore, all corners of wetlands need to be flagged and identified in the field and on the survey map. In addition, this survey may include any other features on the property, such as existing lakes, ponds, creeks, floodplains, other water bodies, and any easements.

A boundary survey is based on meet and bound or bearing and distance measurements. There should be a minimum of two benchmarks on a boundary survey. First, the surveyor needs to ensure that the boundary closes with the minimum standard accuracy to include closure computations or meeting the AHJ requirements, whichever is more stringent.

2.3. Topographic Survey

A topographic survey aims to provide elevations of the site referenced to geodetic survey benchmarks. It is recommended that a contour map be provided based on the slope of the terrain and AHJ requirements. Typically, a one-foot contour map for steep terrain (slopes >forty percent), a two-foot contour map on a medium-sloped terrain (gradients between five percent and forty percent), and a five-foot contour map for flatter terrain are recommended.

A topographic survey may also include a tree survey and the location of existing overground features, such as buildings, structures, etc.

2.4. Tree Survey

A tree survey should be provided where there are numerous trees on site. Most agencies require that trees in diameters greater than six inches at diameter breast height (DBH, diameter measured four feet from existing grade) be included on the survey sheet. However, the exact requirements for the specific jurisdiction should be confirmed. They also can be shown on the landscape plans, depending upon site location, types of trees, environmental needs, and/or AHJ requirements. It is recommended that all identified trees be input in a table or spreadsheet, such as Tables 2.1 and 2.2 shown below. These tables should include all trees, sorted by types and sizes. This way, it is easier to keep track of sick or dead trees, trees to be removed, and finally, trees to be planted to mitigate and compensate for the removed trees.

TABLE 2.1. TREES TO BE REMOVED

Tree Type	No.	Size (inch)	Total (inch)	Remark
Maple	1	7	7	
Pine	2	6	12	
Oak	3	6	18	
Oak	1	7	7	
Oak	1	8	8	
Oak	4	9	36	
Oak	10	2	20	
Oak	11	2	22	
Sub-total			130	Replace 1/3 inch per inch
				130/3 = 44 inches
Live oak	1	12	12	
Live oak	2	14	28	
Live oak	2	25	50	
Sub-total			90	Replace one inch per inch
				90 inches
Total			220	Replacement trees = 44 + 90 = 134 inches

TABLE 2.2 TREES TO REMAIN

Tree Type	No.	Size (inch)	Total (inch)	Remark
Maple	1	21	21	
Pine	1	7	7	
Live oak	37	2	74	
Live oak	3	38	114	
Live oak	4	41	164	
Live oak	1	42	42	
Live oak	3	45	135	
Live oak	2	56	112	
Total			669	

Trees to remain = 669 inches > Trees to be removed = 220 inches

NOTE: This is an example to illustrate tree counts and replace those trees to be removed. Each agency's rules and requirements may differ from others. It is recommended to contact and coordinate with your local AHJ to obtain specific standards and requirements. For example, some agencies may require all landscape plans to be designed and certified by a landscape architect licensed in the state for the work to be performed.

2.5. Critical Area Survey

A critical area survey is required when the subject property is located and covers steep slopes, wetlands, or floodplains. Each of these areas has specific requirements to be surveyed. In addition, most often have strict requirements to mitigate and compensate for any impact on these areas.

For projects that impact steep or erosion control areas, the engineer and preferably a geotechnical engineer may be required to design and certify retaining walls, pond design, or slope stabilization. In addition, geotechnical engineers may propose other methods to ensure that, eventually, there will be no adverse impact on other adjoining property or structures, such as specific stormwater measures or setbacks.

For wetlands impact, the engineer, preferably an environmental engineer, must provide a mitigation plan to compensate for any impact on the wetland. This mitigation plan also depends on the quality of the wetland to be impacted. Although federal laws are to be followed, each state and/or local jurisdiction may have its own more restricted criteria. For example, the mitigation ratio may be more than one-to-one. For some jurisdictions, up to three-to-one with significantly restrictive monitoring plans must be followed for three years. In addition to the area and species mitigation, many agencies may require that the hydroperiod fluctuations within the wetland be maintained depending on the wetland quality and sensitivity.

For floodplain impact, the mitigation is mainly related to the volume compensation.

2.6. Right-of-Way (ROW) Survey

ROW survey is often performed by the survey crews working for the various government agencies such as state, county, or cities. Private sector surveyors conduct a ROW survey on behalf of the agencies mentioned above or their clients as their properties front the ROW. ROW information obtained by the surveyors is usually reflected in the GIS maps maintained by the AHJ and provided to the developers and engineers as a supplement to their project data.

Regardless, it is an excellent practice to contact the GIS department of the AHJ and obtain as much information as available about the ROW fronting a project site (See Chapter 5 for ROW Information and Classification).

2.7. Construction Layout Survey

A construction layout survey is one of the most crucial construction activities that requires accuracy and patience. It is recommended that the contractor should not merely rely on the survey crew for accuracy. A competent contractor will review the field survey layout against construction plans to make sure, at least visually, there are no discrepancies between the field layout and construction plans. The EOR may get involved to check construction layout on large construction sites.

The same professional surveyor responsible for the original survey, such as boundary, topographic, and tree surveys, is preferred to perform the construction layout surveys. Since they are already familiar with the site condition, it probably saves time and money for the contractor or the owner. However, that may not always be possible. Some contractors have a surveying department or are affiliated with surveyors and prefer that their survey staff perform the construction layout survey.

After all, the survey stakes set the stage for the rest of the construction. Therefore, any error may carry on throughout the construction and magnify each step.

2.8. As Built or Record Drawing Survey

Many permitting agencies require an as-built (also called record drawing) survey as a condition of the permit approval of the developer's project within their jurisdiction. This survey occurs after the construction completion of the subject project. In some cases, AHJ may require the as-built survey before issuing the certificate of occupancy. This requirement puts enormous pressure on the contractor and the engineer preparing and submitting the as-built survey.

It is customary for the general contractor to note any significant changes during construction on a set of final approved prints and provide those prints to the EOR to implement in the AutoCAD drawing, MicroStation, or other acceptable formats to be submitted to the AHJ. This record must be performed correctly, professionally, and organized to avoid the entire project's survey.

2.9. Other Surveys

Many other surveying activities are available, with very little or no applicability in civil and site engineering design. They include, but are not limited to, the bathymetric survey (survey showing the depth of bodies of water such as a lake), monitoring survey (survey showing any movement or lack thereof on any structure or point such as train movement or movement due to an earthquake) and aerial photogrammetry surveys.

2.10. Boundary and Topographic Survey Minimum Standards

The following minimum standards provide basic guidelines for submitting Boundary and Topographic surveys to the permitting agencies.

1. A professional land surveyor registered in the state where they perform the work must prepare the boundary and topographic survey.

2. The ground survey and mapping shall cover the subject site plus an additional twenty-five to fifty feet of the immediately adjoining properties. It should also include the complete pavement section of adjacent streets for fifty to one hundred feet.

3. All land surveys submitted for land use approval shall state on their face the "basis of bearing" and "vertical datum," respectively. Before starting the field survey, these data or any other benchmark requirements must be verified and confirmed by the AHJ. Examples of such datums are the National Geodetic Vertical Datum of 1929 (NGVD1929), the North American Datum of 1983 (NAD83), and The North American Vertical Datum of 1988 (NAVD88).

4. Each lot corner must be field tied to at least two (AHJ approved) survey control network monuments. Topographic elevations shall be referenced to the AHJ approved vertical control benchmarks.

5. Provide vicinity map, north arrow, and graphical scale.

6. Include the legal description of the parcel to be surveyed.

7. Provide sufficient geometry to locate all parcels, lots, tracts, and easements accurately.

8. Recording numbers and a brief description of any easements, maintenance agreements, covenants, and restrictions affecting the subject property.

9. Show all existing structures with their locations dimensioned with measurements perpendicular to the property lines.

10. Locations, size, description (including conditions), rim, and invert elevations of all utility structures. Field locates all visible appurtenances and uses the best available information, such as as-built drawings and GIS map for buried utilities.

11. Show any lakes, rivers, streams, ditches, wetlands, or ponds. Indicate the top and bottom of banks for water bodies and the line on ordinary high water on lakes. Show FEMA a hundred-year floodplain.

12. Show all impervious and pervious surfaces to include the edge of pavement or curbs, curb cuts, sidewalks, wheelchair (ADA) ramps, landscape areas, pedestrian or bike paths, rockeries, retaining walls, fences, bridges, culverts, etc.

13. Spot elevations at twenty-five to fifty feet grids, significant grade brakes, property corners, building envelop and corners, top and bottom of all walls and rockeries, centerline, and edge of the pavement, and back of curb at twenty-five to fifty feet stations on all streets.

14. Indicate contours at five-foot intervals for sites with less than five percent slopes. Contours at two-foot intervals for areas with less than forty percent but greater than five percent slopes. Contours at one-foot intervals for sites with greater than forty percent slopes.

15. Distinguish the site with slopes less than forty percent and greater than forty percent. Identify the top forty percent slope.

16. For commercial, multifamily, short plats, subdivisions, and planned unit development (PUD) projects, identify trees six inches or greater in diameter four feet from the existing grade. Label each tree with the common name and diameter. This requirement may greatly vary from AHJ to AHJ. Therefore, all these must be appropriate for the jurisdiction as measured per their code requirements. If a site is heavily wooded, contact the AHJ to discuss the minimum standards for identifying trees.

17. Identify wetland boundaries as flagged by a wetland biologist, field review and approved by the AHJ expert, and subsequently surveyed.

18. A professional geotechnical engineer or engineering geologist, licensed to practice in the state where the project is located, can flag, and identify critical areas such as landslide hazards.

Please be advised that the above list is not inclusive and can significantly vary based on the size and scope of a particular project. Therefore, it is always good practice to communicate and coordinate various standards and the minimum requirements with the AHJ.

CHAPTER 3: SOIL INVESTIGATION AND REPORT

3.1. Introduction

This chapter discusses the importance of a soil investigation and report before starting design and plan preparation. This investigation is even more critical for structural design, which is outside the scope of this book. For site engineering, like surveying, usually a civil engineer is the most qualified professional to come up for scoping of work for the geotechnical report. They may request services based on the specific site design in mind, applicable agencies' requirements, and other site engineering-related items. However, a structural engineer would be more qualified to develop the scope of work for the structural design of a building and foundation.

What is a geotechnical survey or report?

A geotechnical report entails gathering information about the soil's physical characteristics and rocks that make up the land. Geotechnical engineers and geologists complete the geotechnical report to help with designing various elements of a civil or structural engineering project.

This chapter discusses the importance of soil investigation in determining soil types and classes on a site. What should a developer expect in a geotechnical report as a minimum? Like many other engineering services, each service provider develops its means and methods to provide these services efficiently and timely. Therefore, basic knowledge helps developers look for deficiencies and ensures their development needs are met.

As a minimum, a geotechnical report should include the following sections outlined below.

3.2. Project Location and Description

Nowadays, identifying the project location is somehow basic and easy to accomplish using GIS maps and other available resources. Project description, on the other hand, may not be so simple. This section should include detailed explanations of the geotechnical engineer in charge of the project to explain their understanding of their expectations. It should follow and elaborate on the owners' or their civil engineer's scope of work and ensure everyone is on the same page.

What is Scoping?

Scoping outlines the parameters and limitations of the geotechnical engineer's understanding of the project objectives and their proposal to accomplish them professionally and efficiently. It also outlines a step-by-step procedure to meet the project schedule objective within the expected budget.

In many instances, and particularly for larger projects, geotechnical engineers communicate and coordinate the scope of services by submitting a draft letter of understanding to the civil engineer or the structural engineer in charge to obtain approval before proceeding with the investigation.

3.3. Purpose and Methodologies

This section should explain the purpose of the services and may include, but are not limited to, the following:

Degree and extent of the exploration of the surface and subsurface conditions,

Estimate of the groundwater elevation (GWE) or preferably the seasonal high-water elevation (SHWE),

General design recommendation of the civil engineer responsible for the stormwater management system and retention/detention pond design. Soil permeability or percolation rate through the ground. Geotechnical engineering information regarding pavement sections for parking, drive aisles, etc. A general recommendation for the site preparation.

There are numerous methods and types of equipment for evaluating and exploring soil conditions. Since each site is different, this section also identifies subsurface conditions and investigation strategies.

3.4. Field Explorations and Laboratory Testing

This section should identify the number, depth, and types of borings and each boring location. It should indicate the types of testing and their reference to *American Standards Testing Manual* (ASTM) guidelines. It may refer to any portion of the soil sample carried to the laboratory for visual observation or additional testing.

Sheet 5 of 15 in Chapter 7, Project 2 indicated two Soil Borings Logs, SB-1 and SB-2. Similarly, Sheet 7 of 18 in Chapter 8, Project 3 showed two Soil Borings Logs B-01 and B-02.

A professional geotechnical engineering firm collected and provided the soil profiles. In addition to the ground and SHWE, each of the above soil profiles includes information regarding the soil types and classifications of the soil materials at each layer as it changes through the ground, plus other essential data, such as termination depth. The locations of the borings are also shown on a scaled map, which is generally for the site/civil engineering projects picked at the low area where the detention or retention ponds locations are proposed.

Another essential piece of information on these profiles is the date the boring log started and the date it was completed. Depending on the pond's location, there could be a significant change in a wet season SHWE, and during a dry season. Since the SHWE is critical in a dry pond design, boring should take place during the wet season and not the driest time of the year.

3.5. Findings

This section should identify the surface and subsurface conditions of the project site. It explores the existing data, such as the USDA Soil Conservation (SCS) Soil Survey of the County and other available governmental agencies' sites and data. Comparing the existing surface and subsurface data and the explored data confirms the accuracy of some critical information, such as the SHWE, permeability, and soil types.

The United States Department of Agriculture (USDA) Web Soil Survey (WSS) website has soil maps, data, and other helpful information available to everyone to use and explore. The following three pages are an example of exploring some basic but essential soil surveys obtained from the USDA WSS website.

After opening the site, click on the WSS start bottom. Then, click on the address box and provide your address (no need for the city, state, or zip code). We selected the map of Gary Grant Park in Kent, Washington, on page one. Above the map on the left corner, under the area of interest interactive map, pick one of the two areas of interest (AOI). Create a rectangular AOI on the map. We chose the rectangular button, the second from the right. The selected area will automatically be hatched. Then, we picked the soil map on the left of the AOI on the top, and the soil symbols will be shown on the map. Page two provides the map legend and map information. Page three provides the map unit symbols, names, slopes, acreage, and percentage of each soil. Although this is great information to have at the beginning of a project and for a feasibility study, they do not replace the actual soil survey and field investigation and report.

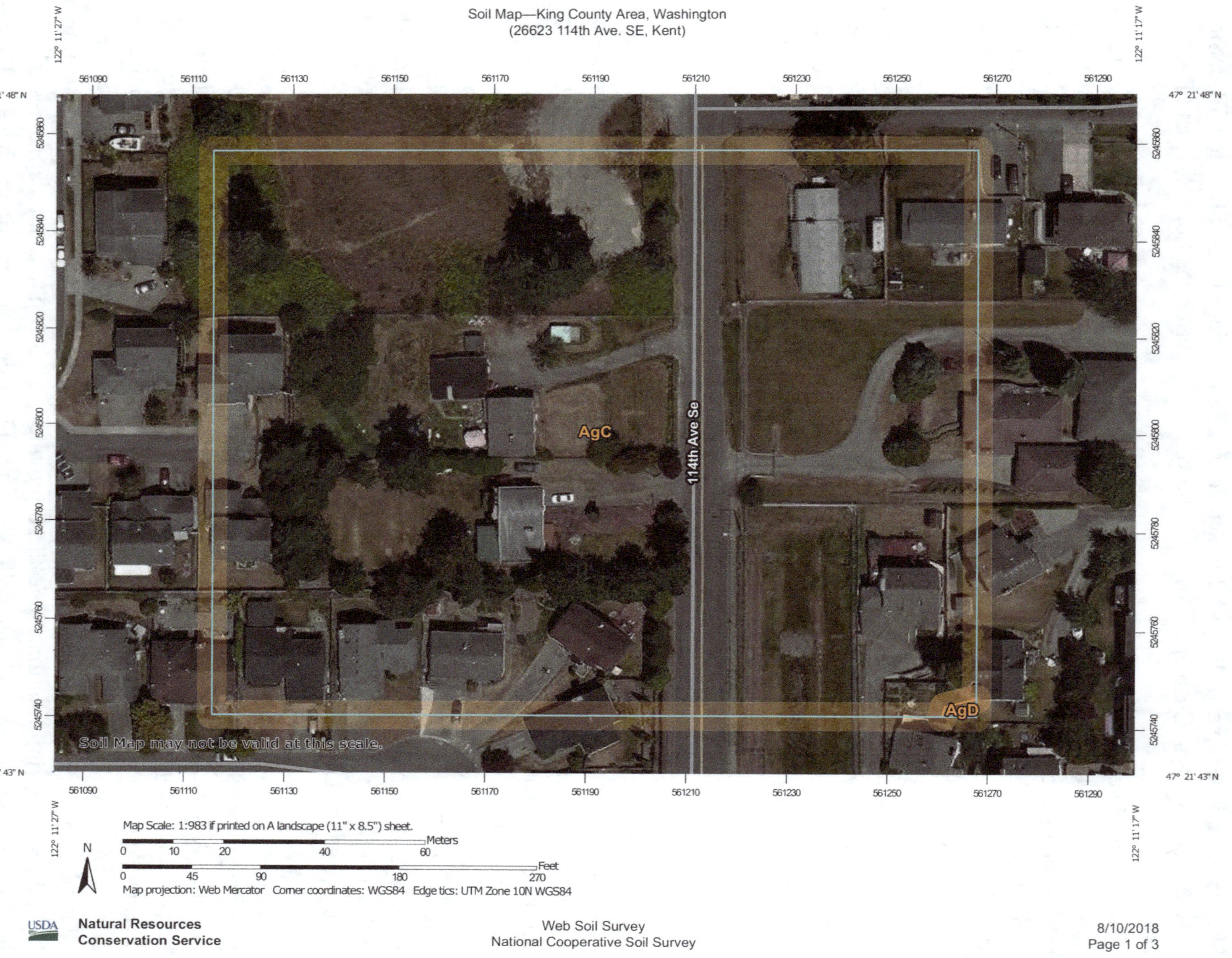

114th Ave Se
AgC
AgD
Soil Map may not be valid at this scale.
Map Scale: 1:983 if printed on A landscape (11" x 8.5") sheet.
Meters
0   10   20   40   60
Feet
0   45   90   180   270
N
Map projection: Web Mercator   Corner coordinates: WGS84   Edge tics: UTM Zone 10N WGS84
USDA   Natural Resources
Conservation Service
Web Soil Survey
National Cooperative Soil Survey
8/10/2018
Page 1 of 3

SITE DESIGN & ENGINEERING

MAP LEGEND

MAP INFORMATION

Area of Interest (AOI)

Area of Interest (AOI)

Soils

Soil Map Unit Polygons

Soil Map Unit Lines

Soil Map Unit Points

Special Point Features

Blowout

Borrow Pit

Clay Spot

Closed Depression

Gravel Pit

Gravelly Spot

Landfill

Lava Flow

Marsh or swamp

Mine or Quarry

Miscellaneous Water

Perennial Water

Rock Outcrop

Saline Spot

Sandy Spot

Severely Eroded Spot

Sinkhole

Slide or Slip

Sodic Spot

Spoil Area

Stony Spot

Very Stony Spot

Wet Spot

Other

Special Line Features

Water Features

Streams and Canals

Transportation

Rails

Interstate Highways

US Routes

Major Roads

Local Roads

Background

Aerial Photography

The soil surveys that comprise your AOI were mapped at 1:24,000.

Warning: Soil Map may not be valid at this scale.

Enlargement of maps beyond the scale of mapping can cause misunderstanding of the detail of mapping and accuracy of soil line placement. The maps do not show the small areas of contrasting soils that could have been shown at a more detailed scale.

Please rely on the bar scale on each map sheet for map measurements.

Source of Map: Natural Resources Conservation Service
Web Soil Survey URL:
Coordinate System: Web Mercator (EPSG:3857)

Maps from the Web Soil Survey are based on the Web Mercator projection, which preserves direction and shape but distorts distance and area. A projection that preserves area, such as the Albers equal-area conic projection, should be used if more accurate calculations of distance or area are required.

This product is generated from the USDA-NRCS certified data as of the version date(s) listed below.

Soil Survey Area: King County Area, Washington
Survey Area Data: Version 13, Sep 7, 2017

Soil map units are labeled (as space allows) for map scales 1:50,000 or larger.

Date(s) aerial images were photographed: Jul 8, 2014—Jul 15, 2014

The orthophoto or other base map on which the soil lines were compiled and digitized probably differs from the background imagery displayed on these maps. As a result, some minor shifting of map unit boundaries may be evident.

Map Unit Legend

Map Unit Symbol	Map Unit Name	Acres in AOI	Percent of AOI
AgC	Alderwood gravelly sandy loam, 8 to 15 percent slopes	4.4	99.9%
AgD	Alderwood gravelly sandy loam, 15 to 30 percent slopes	0.0	0.1%
Totals for Area of Interest		**4.4**	**100.0%**

3.7. Recommendations

This section is probably the most crucial portion of the soil investigation and report. It requires the geotechnical engineering firm to deliver to their client the conclusion of what they performed during their field exploration and laboratory testing. Of course, depending on the scope of services and the complexity of a project, this section can be pervasive. Below are some of the more common items that are usually recommended for a site improvement project.

Groundwater and the SHWE

As mentioned before, the SHWE and GWE estimates are essential. The SHWE is needed to design a detention or retention system properly, and it is a required parameter by most AHJs. When designing dry detention or a retention pond, it is essential the SHWE be at least two feet below the bottom of the pond to allow percolation through the soil. Also, it ensures the pond remains dry after back-to-back storm events. The GWE is helpful when dewatering is necessary. The contractor may decide to perform the work during the most favorable season for dewatering or construction without needing to dewater.

Excavation - Clearing and Rough Grading

Although managing the site during excavation to prevent sediment-laden runoff from being discharged off-site is an essential responsibility of the EOR during design and the contractor during the construction, the geotechnical engineer may recommend construction entrance specifications and temporary erosion and sedimentation controls (TESCs), such as silt fences, and where they should be installed for the best-optimized result. Chapter 7, Project 2 discusses the TESC subject in under Best Management Practices Plans and Details.

Many AHJ require a certified erosion and sediment control lead (CESCL) to have input during the design and construction. Their responsibility is to routinely inspect the site during construction and after a storm event to ensure the site's temporary BMP is always maintained. Still, any input from the geotechnical engineer regarding any particular attention to grades, soil type, sensitivity, materials, and technique is welcome.

Percolation or Permeability Rate

Suppose the subsurface soil investigation results conclude that the soils are permeable, and infiltration is feasible at the site. In that case, the geotechnical engineer may proceed to obtain and test the permeability rate.

Percolation or permeability rate is only required if the site is believed to be suitable for a dry detention system. For example, shallow glacial till site soils are relatively impermeable and not feasible for infiltration. On the other hand, sandy soils investigated in Chapter 7, Project 2, are suitable for infiltration. This value is not needed if the EOR, based on experience, determines that a wet detention pond or some other type of system, such as a concrete vault, is to be designed and constructed.

As part of the geotechnical investigation and report, the geotechnical engineer may test the soil permeability rate and compare it with the available soil data. The soil permeability rate is also available on county or other governmental agencies' resources.

In Chapter 7, Project 2, according to the Natural Resources Conservation Service, soil type Seffner Fine Sand, the permeability rate of 6.0 to 20.0 inches per hour for up to eighty inches in depth. The average permeability rate used for Project 2 was equal to thirteen inches per hour, or (20+6)/2.

The geotechnical engineer may also recommend the site is or is not feasible for other best management practices based on the soil permeability test performed.

Structural Fill

Structural fill is defined as soil fill materials supporting building foundations, floor slabs, pavements, sidewalks, or other related work. Structural fill should be free of organic and other harmful substances and have a maximum fragment size specified by the geotechnical engineer. This section requires input and advice from the geotechnical engineer, especially for the building foundation, piling, steep slopes, soil nails, and retaining walls.

The geotechnical engineer may allow the use of materials, such as recycled crushed concrete or crushed rock, in some regions of the project depending on compaction requirements and other criteria.

Utility companies and AHJ may have their requirements regarding fill and compaction requirements regardless of the geotechnical engineer's recommendation. Sometimes, the property owner or the engineer may request a waiver from the needs of more stringent criteria imposed by the AHJ.

Pavement Sections

Depending on the type of pavement, rigid (concrete) or flexible (asphalt) pavement section, the geotechnical engineer may recommend and provide a specification for a typical pavement section. It also depends on the type of vehicular use for the subject area. A rigid pavement or thicker section may be proposed for high use and heavy vehicular traffic areas, such as semi-trailers, versus everyday use, such as passenger vehicles.

Regardless of the pavement section the engineer or the geotechnical engineer proposes, many AHJ have developed their pavement sections for various uses. The agencies may have firsthand knowledge and experiences with the local soil capability and materials to create a typical roadway or street section that best fits their jurisdiction. It is essential to communicate and coordinate with the AHJ before finalizing a standard pavement section.

Some AHJ may assign a structural number to their standard section and allow the EOR to replace some materials within their typical prescript section, as long as the pavement structural number meets and preferably exceeds the standard structural number. It is probably the best way to compromise if there is a shortage of certain materials during the construction.

For example, the City of Lakeland in Florida stipulates the minimum structural number allowable for any residential roadway flexible pavement section design for their local residential streets shall be 2.59. They also recommend the minimum structural number of a residential collector be 3.06, and the minimum structural number for a street constructed in a commercial or industrial zoned area is 3.28.

The following calculations present how to compute the structural number associated with a local street and a commercial one. The above numbers are based on the layer coefficient assigned to different layers/materials, as indicated in the *Florida Department of Transportation (FDOT) Flexible Pavement Design Manual.* Table 5.4 is shown on the next page, followed by the City of Lakeland Typical Roadway Section.

TABLE 5.4

STRUCTURAL COEFFICIENTS FOR DIFFERENT PAVEMENT LAYERS

Group	Layer Type	Layer Coef. Per unit Thickness	Spec. Sect.
Friction Courses	FC-5	0.00	337
	FC-12.5, FC-9.5	0.44	337
Structural Courses	Superpave Type SP (SP-9.5, SP-12.5, SP-19.0)	0.44	334
Base Courses (General use)	Limerock (LBR 100)	0.18	200
	Cemented Coquina (LBR 100)	0.18	250
	Shell Rock (LBR 100)	0.18	250
	Bank Run Shell (LBR 100)	0.18	250
	Graded Aggregate (LBR 100)	0.15	204
	Type B-12.5	0.30	280
Base Courses (Limited use)	Limerock Stab. (LBR 70)	0.12	230
	Shell Stab. (LBR 70)	0.10	260
	Sand Clay (LBR 75)	0.12	240
	Soil Cement (500 psi)	0.20	270
	Soil Cement (300 psi)	0.15	270
Stabilization	Type B Stab. (LBR 40)	0.08	160-2
	Type B Stab. (LBR 30)	0.06	160-2
	Type C Stab.	0.06	160-2
Subgrade	Cement Treated (300 psi)	0.12	170
	Lime Treated	0.08	165

Page 5.14.0

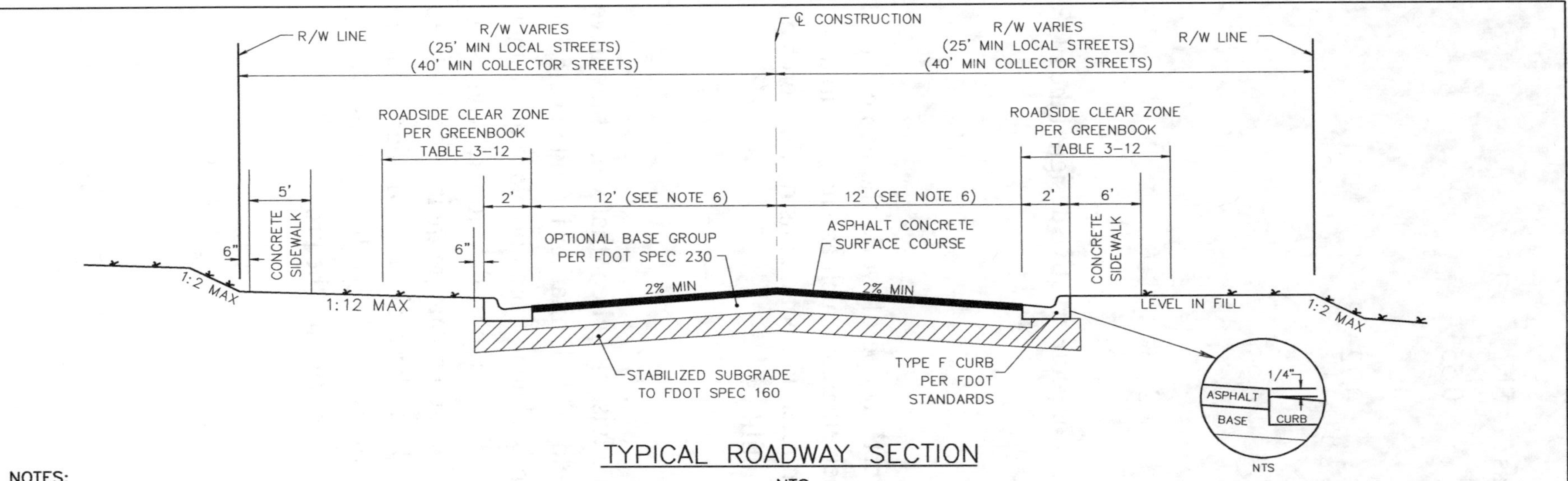

TYPICAL ROADWAY SECTION
NTS

NOTES:

1. SURFACE COURSE OF ALL ROADWAYS SHALL BE ASPHALT CONCRETE MEETING THE FDOT PAVEMENT DESIGN MANUAL AND STANDARD SPECIFICATIONS FOR ROAD & BRIDGE CONSTRUCTION. A MINIMUM OF 1-1/4" THICKNESS OF ASPHALT CONCRETE SHALL BE USED ON LOCAL RESIDENTIAL STREETS AND 1-1/2" OF ASPHALT CONCRETE ON RESIDENTIAL COLLECTOR STREETS. FOR PROJECTED TRAFFIC VOLUMES WITH AN AADT AT OR ABOVE 7500 VPD, OR PROJECTED TRUCK VOLUME AT OR ABOVE 5%, A PAVEMENT STRUCTURE FOR THE INTENDED USE SHALL BE DESIGNED, IN ACCORDANCE TO THE FDOT FLEXIBLE PAVEMENT DESIGN MANUAL.

2. BASE COURSE TO BE COMPACTED LIMEROCK MEETING CURRENT FDOT STANDARDS SPECIFICATIONS FOR ROAD & BRIDGE CONSTRUCTION. A MINIMUM THICKNESS OF 8" SHALL BE USED ON ALL COLLECTOR ROADS. A MINIMUM THICKNESS OF 6" SHALL BE USED ON LOCAL RESIDENTIAL STREETS. A GEOTECHNICAL REPORT IS REQUIRED

3. STABILIZED SUBGRADE TO MEET CURRENT FDOT STANDARD SPECIFICATIONS FOR ROAD & BRIDGE CONSTRUCTION, HAVING A MINIMUM THICKNESS OF 12" ON LOCAL AND COLLECTOR STREETS.

4. LIMEROCK BASE AND STABILIZED SUBGRADE SHALL BE COMPACTED TO 98% OF MAXIMUM DENSITY AS DETERMINED BY AASHTO T-180. THE CITY'S INSPECTOR MAY REQUEST ADDITIONAL DENSITY TESTS IN ADDITION TO THE STANDARD TESTING LOCATIONS. IN ADDITION THE INSPECTOR MAY REQUEST ANY MATERIAL THAT IS NOT FROM AN FDOT APPROVED SOURCE BE SUBJECT TO ADDITIONAL PROCTORS TAKEN FOR EACH LOAD.

5. CONCRETE FOR CURBS SHALL MEET FDOT SPECIFICATIONS FOR CLASS I CONCRETE.

6. RESIDENTIAL LOCAL ROADS MAY BE REDUCED TO 10 FT. TRAVEL LANES AND RESIDENTIAL COLLECTOR ROADS MAY BE REDUCED TO 11 FT. LANES, IF APPROVED IN WRITING BY THE DIRECTOR OF PUBLIC WORKS.

7. ADDITIONAL INFORMATION REGARDING TYPICAL SECTIONS MAY BE FOUND IN THE LAND DEVELOPMENT CODE.

8. SIDEWALK TO BE CONSTRUCTED PER THE LAND DEVELOPMENT CODE.

CITY OF LAKELAND FLORIDA		STANDARD DETAIL	
TYPICAL SECTION WITH CURB & GUTTER			
DATE	SCALE	SHEET NO.	INDEX NO.
REVISED 6/14/18			
REVISED 1/6/16	N.T.S.	1 OF 9	101
REVISED 1/23/15			
REVISED 12/14/00			

For a local residential street:

Layer/material	Coefficient	Min. thickness	SN
Asphalt concrete Surface course	0.44	1 1/4"	0.55
Optional base group Per FOT Spec. 230	0.12	6"	0.72
Stabilized subgrade to FDOT spec. 180	0.08	12"	0.96
Total			2.23

NOTE: This value is lower than 2.59, leaving room for a more conservative design if the EOR decides to use different materials approved by the AHJ.

Other Elements

Depending on the size and the complexity of the project's scope of work, the geotechnical engineer report may include, but is not limited to, other elements such as site preparations, excavation methods, construction entrance(s), detention/retention pond construction, erosion and sediment control, erosion and seismic hazards, and utilities.

Limitations

Finally, every soil investigation report has a section to advise the client that what they have investigated is limited to what they explored in the field. This limitation is understandable since soil can be tricky, and the soil strata can significantly change from one place to another. They do not accept responsibility and liability for any information or data found even a few feet away from the boring locations. In addition, they like to encourage their client to keep the services of the geotechnical engineering firm throughout the project, particularly during excavation, foundation, piling design, and construction. In other words, whenever the soil investigation and the geotechnical engineer input become necessary for the project.

CHAPTER 4: DESIGN DEVELOPMENT

4.1. Introduction

This chapter discusses how a designer should go through establishing design and preparing a sound set of construction plans. One of the first steps in designing and constructing any facility is to scope the project and understand the objective. The facility's owner or end-user is the best person to help scope the project. However, in many instances, the owner and the end-user may not be the same. Since the scope and objective can be broad, we will discuss this in more detail as we go through design examples in the following chapters 6, 7, 8, and 9. After defining the project's scope, the designer can prepare a preliminary or concept plan to share with stakeholders, including the AHJ. This process may be repeated until a firm initial layout accepted by all stakeholders is obtained.

4.2. Concept Site Plan

For more extensive projects, a concept plan may be necessary before preparing a preliminary plan. A preliminary survey plan may suffice to start this task. The survey plan should separately show the existing features, such as buildings, wetlands, utilities, septic systems, and water wells. Other existing features, such as parking areas, exiting water bodies, steep slopes, wetlands, majestic trees, wooded areas, etc., that are proposed to remain would also be necessary.

After adding required setback lines, wetland buffers, etc., what is left would be the area to be used for the proposed project. The proposed building, parking area, access drives, ponds, etc. can be shown as bubbles or clouded areas in a concept plan to be shared with the stakeholders. On occasions, a concept plan may not even have a survey plan. If the purpose of this concept plan is to evaluate the property to see if it is worth investing in, the buyer may not want to spend time and money to get a survey map. In these instances, the design professional may use whatever information is available from the AHJ, and other sources such as GIS and Google maps, to prepare a preliminary concept plan to set the stage for the next step.

4.3. Preliminary Site Plan

A preliminary site plan is like a concept plan and uses the same survey sheet, except for providing more accurate dimensions and data. After indicating all the above-mentioned existing features and the required minimum setback and needed wetland and critical area buffers, one can move to the next step and lay out the proposed buildings and parking area with approximate dimensions, set aside ponds area, etc.

A good set of the preliminary plan should be realistic. Although preliminary, drawings should be to scale with dimensions of setbacks, proposed building envelopes, stormwater management systems, parking layout, driveway locations, etc. It doesn't have to provide details regarding any of the mentioned areas, but the prominent occupiers and scope should be defined.

4.4. Preliminary Drainage Plans

As mentioned in Chapter 7, Small Residential Subdivision, the stormwater management system (SWMS) or drainage plan, and all elements within the SWMS are large compared to other utilities and are probably the most land and time-consuming. It should come after preparing a preliminary site plan. The preliminary site plan should include setbacks, landscape buffers, and other critical area buffers before proceeding with the building envelope and other features, such as proposed parking spaces, egress and ingress, driveway and driveway isles, and dumpster location.

An open detention or retention pond can encroach on the setback area. However, this needs to be confirmed with the AHJ. If a vault is used, it is usually placed under the parking area and away from the setback area and landscape buffer.

After laying out the building envelope, parking, etc., the inlets and storm pipes can be laid out, directing the stormwater runoff to the proposed storage facilities such as ponds or vaults. Of course, one could adjust the storm sewer and the storage facility size and dimensions after performing computations and/or running drainage software to ensure meeting the applicable water quality and quantity requirements.

4.5. Preliminary Utility Plans

There are dozens to hundreds of utilities companies in each state in the United States of America, such as power, water, wastewater, cable, and wireless. Although many of these companies may be the same, operating under different identification, it still adds up to thousands of utility companies operating in America. Depending on the project's location, size, and complexity, it may be necessary to contact and coordinate with multiple utility companies for a project. This is, therefore, another critical factor to consider before securing a site is utilities.

Walking on and around a site reveals many essential utilities near the property. Look for stormwater structures for stormwater emergencies or overflow connection discharging from the site, fire hydrant and water valves to provide potable and fire water services to the property, and sanitary sewer manholes to handle wastewater discharge generated by the proposed site.

Other utilities, such as electrical power, gas, and cables, are essential to any development. Looking for power poles carrying electric lines, other cable lines, and gas markers can at least initially ensure these utilities are available.

A preliminary utility plan can be prepared based on the current information obtained from the local agencies' GIS information. This information can also be obtained by contacting the utility companies directly, attending a pre-application meeting, etc. After drawing all current utility information on a preliminary plan, ensure there is no conflict between the existing and proposed utilities with the proposed stormwater management system. This preliminary plan can be shared with all stakeholders and notify the utility companies for additional or more accurate information to detect any potential conflict.

4.6. Water and Sewer Separation

One critical criterion to remember when designing water and/or sanitary sewerage systems is the separation requirements between the two disciplines. *The Ten-States-Standards* is a reliable and informative document source about sanitary sewerage pipes and pump stations that provide required distances between water lines, sewer lines, and other sanitary structures. The member states and provinces include Illinois, Indiana, Iowa, Michigan, Minnesota, Missouri, New York, Ohio, Ontario, Pennsylvania, and Wisconsin. *The TenStates-Standards* recommend the following standards regarding separating wastewater and water lines.

The document is much longer, but I wanted to focus on separation with some explanation. *The Ten-States-Standards* states that the sanitary sewer lines should be at least ten feet from any existing or proposed water main horizontally and measured from the edge to the edge. For gravity sewers, where it is not practical to maintain a ten-foot separation, the appropriate reviewing agency may allow deviation on a case-by-case basis, if supported by data from the design engineer. Such departure may allow installation of the gravity sewer closer to a water main, provided that the water main is in a separate trench or on an undisturbed earth shelf located on one side of the gravity sewer and at an elevation. Hence, the bottom of the water main is at least eighteen inches above the top of the sewer.

Sewers crossing water mains should provide a minimum vertical distance of eighteen inches between the outside of the water main and the outside edge of the sewer, if the water main is either above or below the sewer. The crossing should have sewer joints that are equidistant and as far as possible from the water main's joints. Where a water main crosses under a sewer, adequate structural support is required for the sewer to maintain line and grade.

The water main or the sewer line may be encased if it is impossible to obtain proper horizontal and vertical separation, as stipulated above. The watertight carrier pipe must extend ten feet on both sides of the crossing, measured perpendicular to the water main. The regulatory agency for use in water main construction approves the materials for the carrier pipe.

4.7. Utility Notification Letters

The utility notification letter is perhaps one of the essential elements within the civil engineering design that some designers and engineers sometimes overlook during the design process. A utility notification letter intends to communicate and inform the utility companies in and around the project that there is a potential that the proposed project will impact their utility. This task should happen at a very early stage of the design to allow utility companies to digest the changes and prepare for any redressing and possible modification of their existing utilities.

On many occasions, a high cost may be associated with the utility relocation, causing a dispute between the project's owner and the utility companies, and sometimes involves the AHJ. Regardless, it is a safe and sound practice to give a heads-up to the utility companies by providing them with a preliminary layout along with the project's scope, schedule, etc., as early as possible. In this way, if there is any disagreement, there will be plenty of time to resolve outstanding issues without impacting the project schedule and, hopefully, the cost.

The following page is a sample of a simple utility notification letter that can be utilized for various projects. Of course, the same information can be conveyed via e-mail with the digital drawings and other associated documents to expedite the process.

DATE

CONTACT NAME
COMPANY NAME
ADDRESS

SUBJECT: **PROJECT'S TITLE / NUMBER**
PROJECT LOCATION
SECTION/TOWNSHIP/RANGE and FOLIO or PARCEL NUMBER
COUNTY

Dear NAME:

As shown in the enclosed construction plans, we propose constructing a PROJECT SCOPE at the above-referenced location.

The proposed construction will start upon receiving all necessary permits or on START DATE. Please identify and locate any facilities you may have in the vicinity which may be affected by this construction.

If there are no facilities, please respond in kind. If we have not received a response by DATE, we assume you have no facilities within or around the project area and are not interested in opposing this construction.

If there are questions or concerns regarding this matter, please do not hesitate to contact our office at PHONE NUMBER or EMAIL ADDRESS.

Truly yours,

YOUR NAME

XC: File of Record

4.8. Stormwater Management or Technical Information Report (TIR)

The stormwater management design and plan preparation for site engineering is the most time-consuming task, as we will discuss in chapters 7, 8, and 9. After preparing a preliminary paving, grading, and drainage plan to include pavement spot elevations, inlets, storm sewer pipes, and storage facilities with discharge or control structures, it is time for the designer or the EOR to perform drainage computations.

Although personal computers came into use by the public and universities around 1980, most drainage software available today was unavailable. Before personal computers became popular, designers performed drainage computations manually using some advanced calculators, such as Hewlett Packard (HP). One or two decades later, a variety of drainage software with graphic capability became available and has become more efficient and sophisticated since then.

Today's drainage computer software performs highly complicated computations in less than an hour, going through multiple iterations to provide output data and/or error messages that can be fixed by redesigning and resizing the stormwater management system to achieve an optimal design. Much of this software has built-in rainfall events, geometric arrangements for various storage facilities, control structures, and other relevant data for the designer to select to expedite the design and facilitate the most efficient method.

The designer prepares a drainage report or technical information report (TIR) to submit to the AHJ, along with the final construction plans for agencies' review, approval, and issuance of construction permits after obtaining a satisfying input/output result.

4.9. Covenants and Easements

Like any other utilities, stormwater management systems are not exempt from maintenance. If anything, they probably need more care than many other utilities. The components within a stormwater management system are usually large and consume more areas. They need more manual maintenance and attention. The maintenance requirements are a part of the permit condition for many AHJ. For example, the Southwest Florida Water Management District (SWFWMD) requires the property owner to submit an inspection report certified by a professional engineer or other qualified professional every year or every eighteen months, depending on the system's design. The professional ensures the system is substantially conformed to the originally designed system. If any deficiencies are found during the inspection, the owner must bring the system to compliance. If, for example, the control structure is blown away or missing, it must be rebuilt and recertified to the AHJ.

Drainage easements to access the facilities and other covenants may be needed to provide proper maintenance. Sometimes, the covenant may be for the entire facility, such as the one located in a subdivision, to ensure no encroachment occurs years after the construction is completed and the system starts operating. On occasions, property owners knowingly or unknowingly construct their fences or other structures on access easements and other covenant areas, which causes discomfort for all stakeholders, particularly the maintenance workers. The EOR, usually a civil engineer, provides and designates the necessary covenant and easements for the AHJ review and approval. If accepted by the AHJ, the registered land surveyor responsible for the plat and other survey work usually provides the final legal document. A real estate attorney may also be involved in more extensive and sophisticated projects, such as planned unit developments (PUD) and large subdivisions.

4.10. Final Construction Plans

After attending a pre-application meeting and getting input from all utility companies, and other stakeholders, the designer can prepare and finalize the construction plans as explained in chapters 7, 8, and 9. The process mentioned in those chapters is probably oversimplified and can be much more complicated for more extensive projects, such as large commercial and roadway projects. For those projects, the designer or project manager may repeat this process between the designer, EOR, and various other stakeholders via sharing plans at different stages. The more popular design stage is ten percent, thirty percent, sixty percent, ninety percent, one hundred percent, or could be anywhere between before finalizing the construction plans.

Chapters 7 (Small Residential Subdivision), 8 (Medium Residential Subdivision), and 9 (Commercial Site) cover various projects and how to proceed from vacant land to a final set of construction plans.

These chapters also discuss the pre-development and post-development basin boundaries, which are essential to any SWMS design and plans to meet the AHJ water quality and quantity requirements.

4.11. Homeowners Association (HOA) Document

The HOA document is a legal document that provides instructions and guidance to the homeowners and the legal entity responsible for the maintenance and operation of the entire site.

This document is not mandatory in all states. Although, for larger subdivisions, it is probably an essential document to avoid potential future lawsuits and issues arising from a lack of processes and procedure and/or identification of the entity responsible for maintenance and operation. The HOA document is also legal and should be prepared by a qualified attorney, possibly a real estate attorney. The best time to perform this task is after obtaining all permits, or perhaps after construction completion. However, sometimes the AHJ requires a draft copy of the HOA before issuing permits. That is why the best time to finalize the easements and other covenants is after the construction completion. Easement locations may change due to changes during construction for various utilities. Although, all documents can be prepared in draft format and finalized after construction has been completed to incorporate any adjustments due to construction plans revisions or modifications.

CHAPTER 5: RIGHT-OF-WAY INFORMATION AND CLASSIFICATION

5.1. Introduction

This chapter discusses the meaning of right-of-way (ROW), the ownership, use, and how to access and obtain necessary permits from the agency having jurisdiction (AHJ), and the entity responsible for maintaining the ROW. It also explains different types or classes of ROW available through agencies for an applicant to explore and apply for a ROW permit.

Most agencies in charge of ROW may have five or more applications for their use of the ROW. Some may have an individual category for each group of users, and others may bundle all or some in a master ROW application form. I explain each type, its purpose, and the requirements and challenges for obtaining a particular class of ROW permit.

5.2. What Right-of-Way (ROW) Means?

Public ROW is either on real property or an easement in favor of a public agency (i.e., city, county, state) for public travel, use, and other benefits. They include public streets and properties reserved for public utilities (i.e., sewer/water/storm lines, telephone, power, and other utility infrastructure), transmission lines and extensions, curb and gutter, walkways or sidewalks, bike lanes, and multi-use and equestrian trails. The ROW usually extends ten to twenty feet beyond the roadway pavement for municipalities and can spread much wider for state governments and the Federal government.

5.3. When is a Right-of-Way Permit Required?

What kind of activities may trigger what (if any) ROW permit application submittal to the AHJ for review, approval, and permit issuance? Each jurisdiction has rules for these triggers, and they may sometimes use a different name for them. The city or county's website, where the work takes place, is a good starting- point for specifics. However, knowing when a ROW permit may be required is helpful. In general, it is wise to assume that most work in the ROW will require a ROW permit of some kind and, in some instances, may require more than one permit. If an activity involves the construction and/or some disturbance of the ROW or impacts those who are routinely utilizing it, it is a safe assumption that one or more ROW permits will be required.

Most projects involve at least one ROW permit from the local city, county, or state, except properties without access to a public road (the property accesses a private road before entering a public street).

Does the AHJ over the project site require adding frontage improvements (i.e., curb and gutter, sidewalk, bike lane, landscaped buffer) in front of the development? Is there a need to control traffic during the site construction? Are the site construction activities blocking pedestrian access to the sidewalk? Is there a need for a new sign near the road?

Answering yes to any of these questions will probably trigger the need for a ROW permit. Just like other aspects of permitting, ROW permits are not a standard process. While the exact format and requirements for these permits vary between jurisdictions, there are more commonalities than differences.

This chapter aims to define different types of ROWs, when a certain kind of ROW permit is needed, and how to obtain and secure a ROW permit.

5.4. Types or Classes of Right-of-Way Permits

Some permit agencies arbitrarily designate A, B, C, or Type I, II, III, etc., to their permit applications. Other agencies have a master application form with many check boxes that customers can check based on their required permit activities. Regardless, the critical factor to remember is the use and the purpose of acquiring a specific type or class of ROW permits to using the ROW. We discuss and explain various types or classes of ROW in three categories in the following pages.

5.5. Category I: Short Term, Temporary Use and Non-Disturbance of the Right-of-Way

This category could include, but is not limited to, any of the following activities:

- Assemblies and public gatherings,
- Non-motorized vehicle races, such as bicycle races,
- Block parties,
- Parades,
- Processions,
- Street dances,
- Street runs, and walks.

- Potable On Demand Storage (PODS)
- Filming
- Grand openings

This permit type generally requires a sketch or mockup rather than full plans and detailed traffic control details, plus a narrative. Be aware that there is still a turnaround time for this kind of permit, so it is important to anticipate that one will be required in advance.

Depending on the size of the municipality and the complexity of the government agencies' organization, this list can be much more significant to include various activities. The purpose here is to provide the idea that these activities often require some permits from the AHJ before implementation.

Short term and temporary use usually mean about a week or less, say forty-eight or seventy-two hours for ROW encroachments of a temporary nature. The permit is null and void at the end of this designated period. None-disturbance implies that it is not a construction permit and that no part of the permit allows the applicant to disturb or have any physical impact on the ROW. On occasions, these permit categories may require the installation of traffic control devices to maintain the safe and orderly operation of those participating in the activity. At the end of the construction, traffic control devices must be removed and a cleanup (if necessary) of the ROW to return it to its original condition.

Depending upon the size of the activity and the AHJ, some or all the following conditions may apply to the permit:

- Permit fee must be paid when the permit is issued,
- Some may require a bond to be posted or some liability insurance to indemnify the government agencies and their staff from any liabilities associated with the proposed activities,
- If there is any need for maintenance workers for clean-up, police, or fire departments for safety, etc. the applicant agrees at the time of picking up the permit to pay all invoices incurred during the proposed activities to include any overtime and off-hour works,

For example, in 2015, thousands of Seattle Seahawks fans gathered on S. 188th Street and International Boulevard leading to the SeaTac International Airport in SeaTac to give their favorite team a super send-off. This activity required a Class A permit from the City of SeaTac, which included lane closures, maintenance of traffic (MOT), help from the city staff Public Works Maintenance Division for setting up and cleaning up, and police present to manage the crowd safely. It cost the Seattle Seahawks approximately $45,000 in permit fees and another related cost.

The permit fee can provide a sense of the magnitude of the crowd relative to usual activities, like processions, street dances, street runs, and walks. In comparison, most Class A permit application fees/costs are from a few hundred to a few thousand dollars depending on the activities, location, and the permitting agency.

5.6. Category II: Long Term, Permanent, and Disturbance of the Right-of-Way

This category includes any permanent construction activities within the public ROW. Most permitting agencies issue these types of permits far more than any other, particularly in a booming economy. It may include any of the following construction activities or a combination of:

- Driveway entrances
- Sidewalk
- Curb and gutter
- Irrigation lines
- Potholing
- Trenching
- Jack and bore
- Directional bore
- Illumination

This type of permit will normally need detailed plans, likely prepared by a professional engineer, for all but the most straightforward applications (e.g., potholing). Standard details to look for are permitted options for driveway connections to the ROW—most places only have a handful of allowed options, and approval will be much more complicated if your development does not conform with those. It is essential to research the city or county where the construction is taking place to determine what standard details or notes they provide. The closer you can adhere to these standards in your submittal documents, the faster the permit will be approved. It means the difference between obtaining approval after the first plans and related documents submittal, or getting it after a few rounds of trial-and-error submittal.

This category of ROW permit may also be triggered by municipalities that require frontage improvements, even if that was not the intent of the development. A project must construct the frontage improvements along its property line(s) if it meets specific criteria. The frontage improvements usually involve half-street improvements consisting of travel lane widening, curb and gutter installation, bike lane, parking spaces, landscaping strip, sidewalk, or a combination. Many jurisdictions have typical sections and standard requirements for this purpose. See the following photo for the construction activities within the ROW.

5.7. Franchise Utility Permit

This type of permit can vary widely between jurisdictions. A full-service municipality with its utilities may treat this as a kind of ROW construction permit, so all that will be required is submitting one application and a set of construction plans. In jurisdictions where the utilities are private or not affiliated with the jurisdiction, projects need to be submitted to the utilities who will review them. The applicant addresses any comments or deficiencies before obtaining approval from the outside utility company.

Although utility (i.e., water, wastewater, power, cable) companies fall into this category for their construction activities within the ROW, there is a difference if they choose to have a franchise agreement with the AHJ. Utility companies must abide by their agreement with the AHJ, which may include some advantages. For example, they may postpone paying permit fees when picking up their permit to have the AHJ send them monthly invoices.

In jurisdictions where the utilities are private or not affiliated with the AHJ, construction plans and related documents must be submitted to the utility companies who will review, approve, and issue permits for their respective utilities. Then, the utility companies generally apply for a permit from the AHJ (i.e., city or county). This process may take more time and involve more rounds of revisions to meet the requirements of both the utilities and the city or county. Most cities and counties have a mix of internal utilities and external agencies with franchise agreements for using the ROW. The work may be performed by the contractors or the utility company at the expense of the owner/applicant. If it is the latter, the contractor must schedule a convenient time in the construction schedule with the utility company since most utility companies are busy.

The most important thing to be mindful of in utility design is conflicts. Any permit with digging should include a Call Before You Dig label, which is especially important for utilities. When looking at proposed utilities, it is essential to consider both the horizontal and vertical position of all utilities along the frontage of a property to make sure no proposed installations occupy the same space. A plan sheet should show this information for all utilities, if possible, to make it readily apparent where utilities should be placed relative to each other. There are also rules for how close water and sewer lines can be (generally, a ten-inch minimum) that are important to keep in mind. Also, see Section 4.6, Chapter 4, Design Development, for additional information.

Some utilities may require additional consideration. Many jurisdictions increasingly require or encourage underground electrical and communication connections. That can have a high cost, and it is best to investigate early on if that is necessary for the project, and if the project or needed frontage improvements have any impact on these installations.

Any utility installation in the ROW that blocks part of or is immediately adjacent to a sidewalk or road will need a traffic control plan (TCP). Before designing the utilities, in many cases, the designer will need to pothole or request records from the city, county, or utility for the existing installations to determine where the proposed utility can tie into the system. For more about potholing, refer to Section 5.6. Category II, Long Term, Permanent, and Disturbance of the Right-of-Way mentioned above.

Most jurisdictions have online applications available and may even have fully online submittals for permits. Start with finding the form for the type of ROW permit you seek. Remember that, per the preceding sections, there may be more than one type of form, so it is important to read thoroughly enough to determine the correct application. Fill out that form completely. Taking the time to fill it out right the first time can prevent misunderstandings and needless rounds of permit review. Ensure to provide contact information and a descriptive enough narrative to fully describe the activities permitted under this permit.

Other submittal requirements may include:

- A construction estimates to post a bond for the work (generally a form the jurisdiction provides).
- Payment of permit fees.
- Proof of required business and contractors' licenses.

Another submittal requirement for nearly all types of ROW permits, a TCP, is discussed in more detail below.

5.8. Traffic Control Plans (TCP)

Controlling and properly guiding traffic during civil construction, particularly in the accessed public ROW, is probably one of the most critical aspects of any construction activities. Lack of appropriate control devices throughout the life of construction can cause property damage and even death, leading to an expensive and unsuccessful project outcome. Therefore, it is imperative to pay attention to the details and not cut corners due to the additional expenses of implementing TCP.

There are two ways to develop a sound set of TCPs. If the organization that owns the ROW has standard TCPs for an application like yours, you may be able to submit or reference those plans. Regardless, the best source for TCP is the Federal Highway Administration (FHA) *Manual of Uniform Traffic Control Devices* (MUTCD). It is a good and safe practice to ensure that all referenced and used TCPs conform to the MUTCD.

Various descriptive TCP drawings exist in the *MUTCD* and may be used for a simple construction with limited impact on ROW. There are dozens of similar drawings available from the WSDOT or other state departments of transportation websites that can be utilized for different situations. Most states adopt either the *FHA MUTCD* manual or some modified version of the same manual. The following pages are excerpts from the Washington State Department of Transportation for One-Lane, Two-Way Traffic Control with Flaggers (TC-1), Shoulder Closure – Low Speed (TC-5), Lane Shift-Tree Lane Roadway (TC-12), and Intersection Lane Closure-Tree Lane Roadway (TC-14), respectively. Some municipalities may require modifications to the drawings as mentioned above (i.e., adding a uniformed police officer with a marked car) before approving them for implementation.

A professional engineer or a traffic control supervisor can draft a customized TCP. The TCP should address all types of traffic in the area affected by the work, including vehicular, bike, and pedestrian. Blocking the transit passages, as mentioned above, will require approved signs for the type of closure. A detour route should also be identified if necessary. Almost all jurisdictions require TCPs signed by qualified professionals to obtain approval from the AHJ.

All work zones in the ROW impacting any transit should have a work zone shown on a TCP. Typical components of an acceptable TCP include a title, north arrow, legend, signs with type and size noted, a marked work zone, tables for buffer space, sign spacing, and channelization. Notes should also include any jurisdiction-specific requirements, such as allowable work hours or the presence of an off-duty police officer with a marked police vehicle on significant roadways.

In some jurisdictions or high-traffic areas, the contractor is required to schedule a site inspection for an inspector to inspect and approve the traffic control setup before construction commencement.

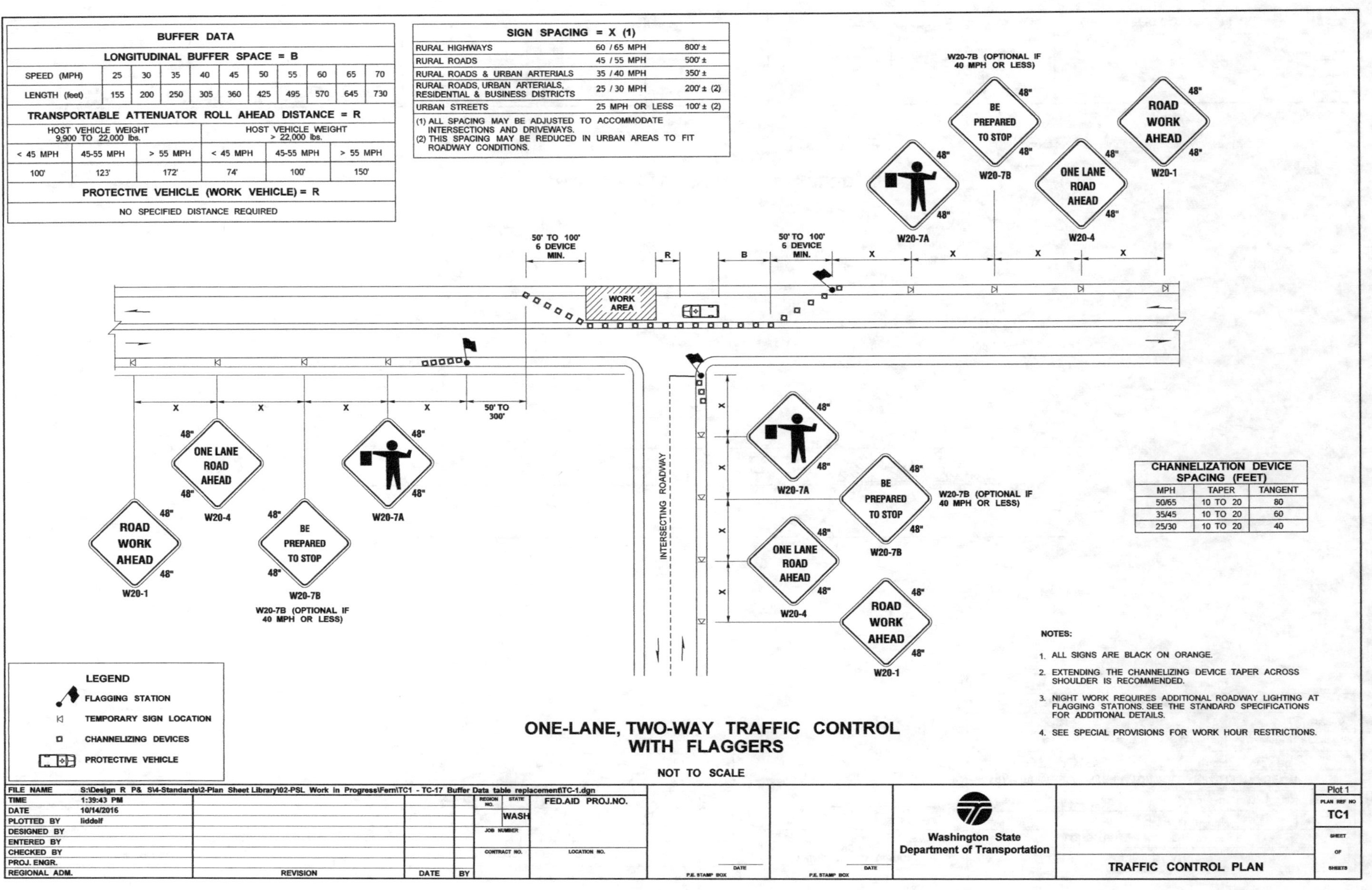
BUFFER DATA
LONGITUDINAL BUFFER SPACE = B
SPEED (MPH) 25 30 35 40 45 50 55 60 65 70
LENGTH (feet) 155 200 250 305 360 425 495 570 645 730
TRANSPORTABLE ATTENUATOR ROLL AHEAD DISTANCE = R
HOST VEHICLE WEIGHT 9,900 TO 22,000 lbs.
HOST VEHICLE WEIGHT > 22,000 lbs.
< 45 MPH 45-55 MPH > 55 MPH < 45 MPH 45-55 MPH > 55 MPH
100' 123' 172' 74' 100' 150'
PROTECTIVE VEHICLE (WORK VEHICLE) = R
NO SPECIFIED DISTANCE REQUIRED

SIGN SPACING = X (1)
RURAL HIGHWAYS 60 / 65 MPH 800' ±
RURAL ROADS 45 / 55 MPH 500' ±
RURAL ROADS & URBAN ARTERIALS 35 / 40 MPH 350' ±
RURAL ROADS, URBAN ARTERIALS, RESIDENTIAL & BUSINESS DISTRICTS 25 / 30 MPH 200' ± (2)
URBAN STREETS 25 MPH OR LESS 100' ± (2)
(1) ALL SPACING MAY BE ADJUSTED TO ACCOMMODATE INTERSECTIONS AND DRIVEWAYS.
(2) THIS SPACING MAY BE REDUCED IN URBAN AREAS TO FIT ROADWAY CONDITIONS.

W20-7B (OPTIONAL IF 40 MPH OR LESS)
BE PREPARED TO STOP
W20-7B
48"
ROAD WORK AHEAD
W20-1
48"
ONE LANE ROAD AHEAD
W20-4
48"
W20-7A
48"

50' TO 100' 6 DEVICE MIN.
R
B
50' TO 100' 6 DEVICE MIN.
X
WORK AREA

X
50' TO 300'
ONE LANE ROAD AHEAD
W20-4
48"
BE PREPARED TO STOP
W20-7B
48"
W20-7B (OPTIONAL IF 40 MPH OR LESS)
ROAD WORK AHEAD
W20-1
48"
W20-7A
48"

INTERSECTING ROADWAY
W20-7A
48"
BE PREPARED TO STOP
W20-7B
48"
W20-7B (OPTIONAL IF 40 MPH OR LESS)
ONE LANE ROAD AHEAD
W20-4
48"
ROAD WORK AHEAD
W20-1
48"

CHANNELIZATION DEVICE SPACING (FEET)
MPH TAPER TANGENT
50/65 10 TO 20 80
35/45 10 TO 20 60
25/30 10 TO 20 40

NOTES:
1. ALL SIGNS ARE BLACK ON ORANGE.
2. EXTENDING THE CHANNELIZING DEVICE TAPER ACROSS SHOULDER IS RECOMMENDED.
3. NIGHT WORK REQUIRES ADDITIONAL ROADWAY LIGHTING AT FLAGGING STATIONS. SEE THE STANDARD SPECIFICATIONS FOR ADDITIONAL DETAILS.
4. SEE SPECIAL PROVISIONS FOR WORK HOUR RESTRICTIONS.

LEGEND
FLAGGING STATION
TEMPORARY SIGN LOCATION
CHANNELIZING DEVICES
PROTECTIVE VEHICLE

ONE-LANE, TWO-WAY TRAFFIC CONTROL WITH FLAGGERS
NOT TO SCALE

FILE NAME S:\Design R P& S\4-Standards\2-Plan Sheet Library\02-PSL Work In Progress\Fern\TC1 - TC-17 Buffer Data table replacement\TC-1.dgn
TIME 1:39:43 PM
DATE 10/14/2016
PLOTTED BY liddolf
DESIGNED BY
ENTERED BY
CHECKED BY
PROJ. ENGR.
REGIONAL ADM.
REVISION DATE BY
REGION NO. STATE WASH FED.AID PROJ.NO.
JOB NUMBER
CONTRACT NO. LOCATION NO.
P.E. STAMP BOX DATE
Washington State Department of Transportation
TRAFFIC CONTROL PLAN
Plot 1
PLAN REF NO TC1
SHEET OF SHEETS

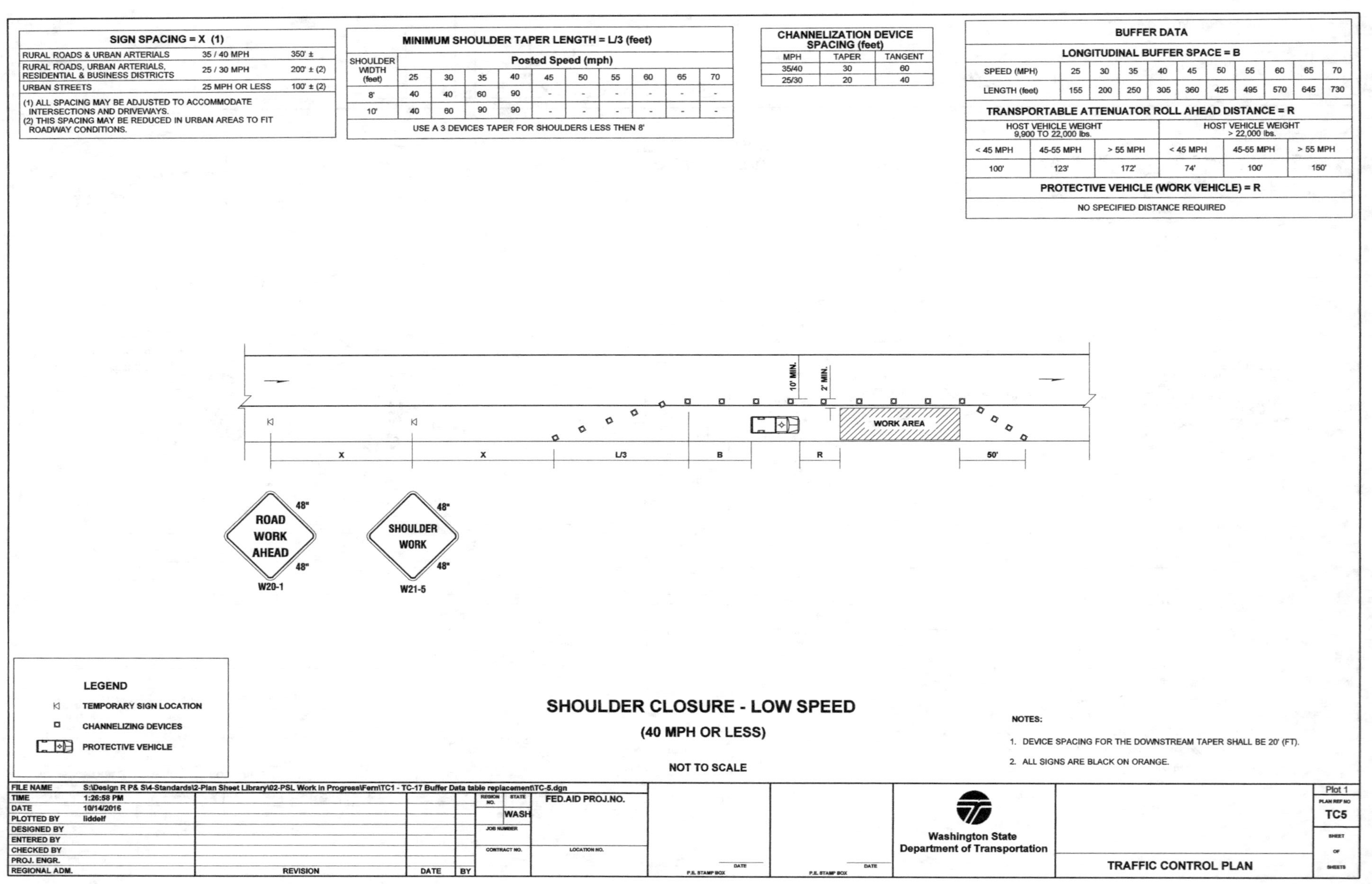

SIGN SPACING = X (1)

RURAL ROADS & URBAN ARTERIALS	35 / 40 MPH	350' ±
RURAL ROADS, URBAN ARTERIALS, RESIDENTIAL & BUSINESS DISTRICTS	25 / 30 MPH	200' ± (2)
URBAN STREETS	25 MPH OR LESS	100' ± (2)

(1) ALL SPACING MAY BE ADJUSTED TO ACCOMMODATE INTERSECTIONS AND DRIVEWAYS.
(2) THIS SPACING MAY BE REDUCED IN URBAN AREAS TO FIT ROADWAY CONDITIONS.

MINIMUM SHOULDER TAPER LENGTH = L/3 (feet)

SHOULDER WIDTH (feet)	Posted Speed (mph)									
	25	30	35	40	45	50	55	60	65	70
8'	40	40	60	90	-	-	-	-	-	-
10'	40	60	90	90	-	-	-	-	-	-

USE A 3 DEVICES TAPER FOR SHOULDERS LESS THEN 8'

CHANNELIZATION DEVICE SPACING (feet)

MPH	TAPER	TANGENT
35/40	30	60
25/30	20	40

BUFFER DATA

LONGITUDINAL BUFFER SPACE = B

SPEED (MPH)	25	30	35	40	45	50	55	60	65	70
LENGTH (feet)	155	200	250	305	360	425	495	570	645	730

TRANSPORTABLE ATTENUATOR ROLL AHEAD DISTANCE = R

HOST VEHICLE WEIGHT 9,900 TO 22,000 lbs.			HOST VEHICLE WEIGHT > 22,000 lbs.		
< 45 MPH	45-55 MPH	> 55 MPH	< 45 MPH	45-55 MPH	> 55 MPH
100'	123'	172'	74'	100'	150'

PROTECTIVE VEHICLE (WORK VEHICLE) = R

NO SPECIFIED DISTANCE REQUIRED

FILE NAME	S:\Design R P& S\4-Standards\2-Plan Sheet Library\02-PSL Work In Progress\Fern\TC1 - TC-17 Buffer Data table replacement\TC-5.dgn
TIME	1:26:58 PM
DATE	10/14/2016
PLOTTED BY	liddelf
DESIGNED BY	
ENTERED BY	
CHECKED BY	
PROJ. ENGR.	
REGIONAL ADM.	

REGION NO.	STATE	FED.AID PROJ.NO.
	WASH	
JOB NUMBER		
CONTRACT NO.	LOCATION NO.	

Plot 1

PLAN REF NO. **TC5**

SHEET / OF / SHEETS

REVISION   DATE   BY

BUFFER DATA

LONGITUDINAL BUFFER SPACE = B

SPEED (MPH)	25	30	35	40	45	50	55	60	65	70
LENGTH (feet)	155	200	250	305	360	425	495	570	645	730

TRANSPORTABLE ATTENUATOR ROLL AHEAD DISTANCE = R

HOST VEHICLE WEIGHT 9,900 TO 22,000 lbs.			HOST VEHICLE WEIGHT > 22,000 lbs.		
< 45 MPH	45-55 MPH	> 55 MPH	< 45 MPH	45-55 MPH	> 55 MPH
100'	123'	172'	74'	100'	150'

PROTECTIVE VEHICLE (WORK VEHICLE) = R

NO SPECIFIED DISTANCE REQUIRED

SIGN SPACING = X (1)

RURAL HIGHWAYS	60 / 65 MPH	800'±
RURAL ROADS	45 / 55 MPH	500'±
RURAL ROADS & URBAN ARTERIALS	35 / 40 MPH	350'±
RURAL ROADS & URBAN ARTERIALS RESIDENTAL & BUSINESS DISTRICTS	25 / 30 MPH	200'± (2)
URBAN STREETS	25 MPH OR LESS	100'± (2)

(1) ALL SPACING MAY BE ADJUSTED TO ACCOMMODATE INTERCHANGE RAMPS, AT-GRADE INTERSECTIONS AND DRIVEWAYS.
(2) THIS SPACING MAY BE REDUCED IN URBAN AREAS TO FIT ROADWAY CONDITIONS.

MINIMUM TAPER LENGTH = L (feet)

LANE WIDTH (feet)	Posted Speed (mph)									
	25	30	35	40	45	50	55	60	65	70
10	105	150	205	270	450	500	550	-	-	-
11	115	165	225	295	495	550	605	660	-	-
12	125	180	245	320	540	600	660	720	780	-

CHANNELIZATION DEVICE SPACING (feet)

MPH	TAPER	TANGENT
50/60	40	80
35/45	30	60
25/30	20	40

PCMS #1

1	2
LEFT LANE CLOSURE	1 MILE AHEAD
2.0 SEC	2.0 SEC

FIELD LOCATE IN ADVANCE OF TEMPORARY SIGNS.

PCMS #2

1	2
LANE SHIFTS LEFT	1 MILE AHEAD
2.0 SEC	2.0 SEC

FIELD LOCATE IN ADVANCE OF TEMPORARY SIGNS.

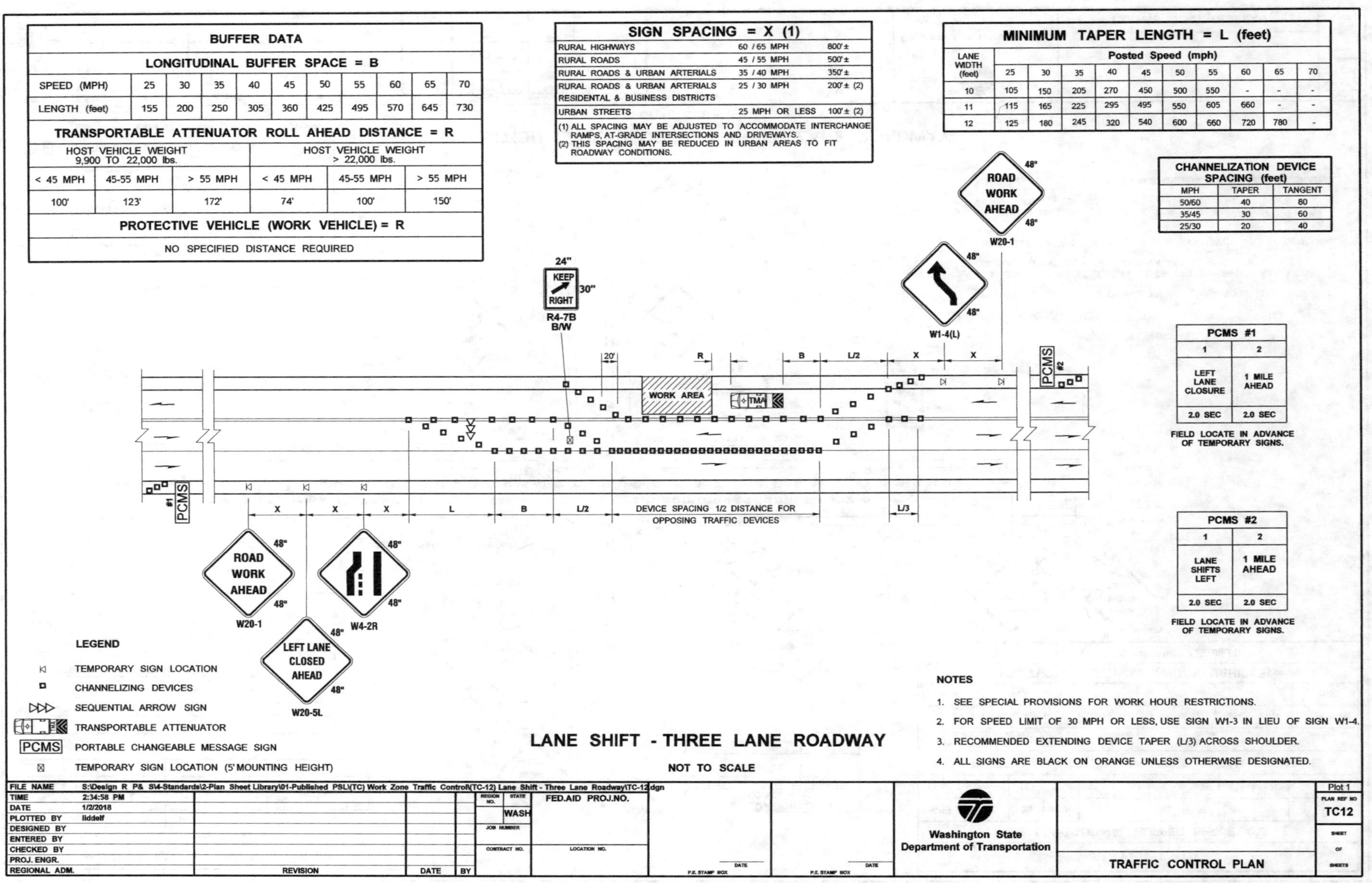

LANE SHIFT - THREE LANE ROADWAY

FILE NAME	S:\Design R P& S\4-Standards\2-Plan Sheet Library\01-Published PSL\(TC) Work Zone Traffic Control\(TC-12) Lane Shift - Three Lane Roadway\TC-12.dgn				Plot 1
TIME	2:34:58 PM				PLAN REF NO TC12
DATE	1/2/2018				
PLOTTED BY	lldself				
DESIGNED BY					
ENTERED BY					
CHECKED BY					
PROJ. ENGR.					
REGIONAL ADM.	REVISION	DATE	BY		

FED.AID PROJ.NO.

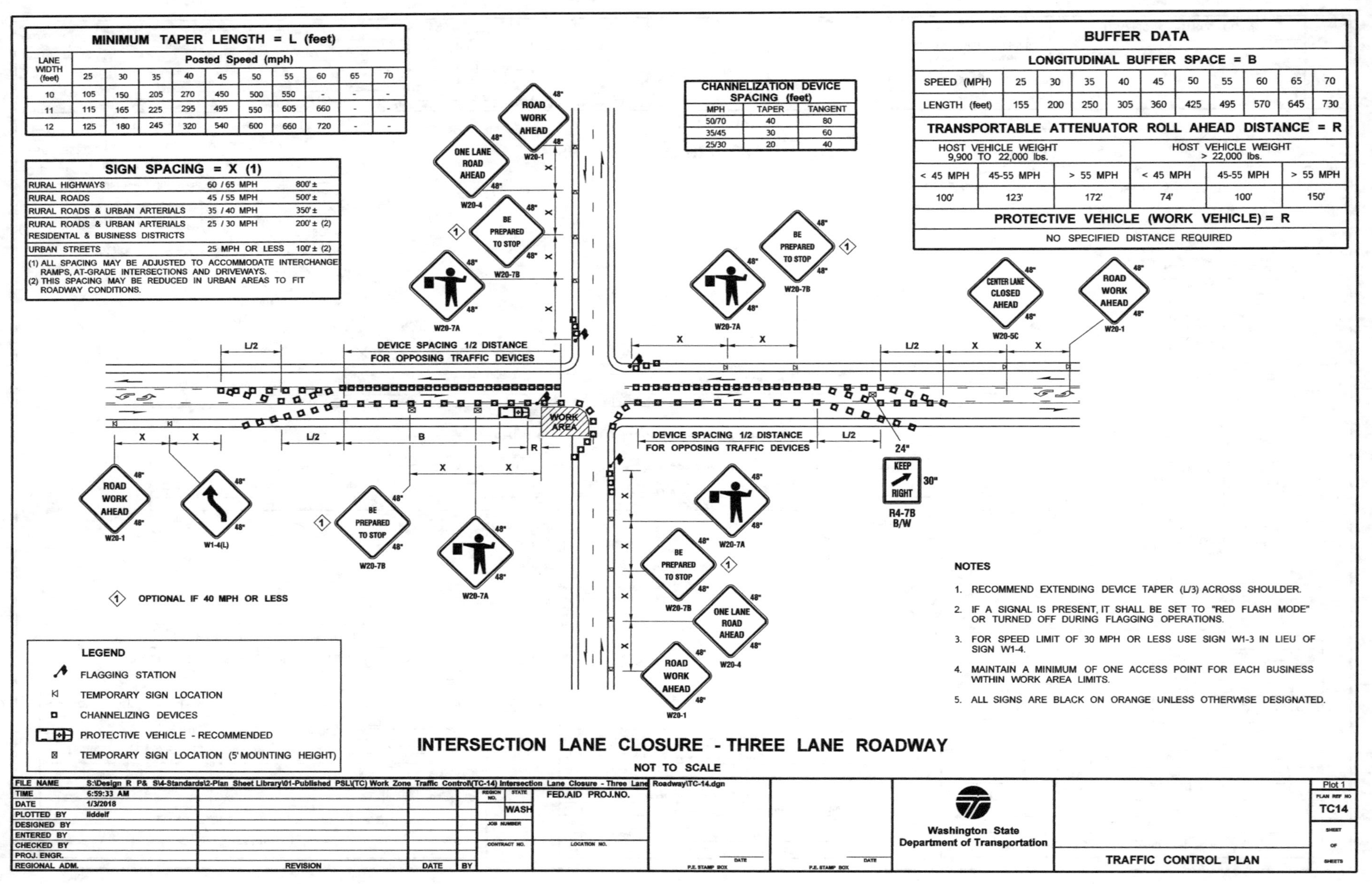

MINIMUM TAPER LENGTH = L (feet)
LANE WIDTH (feet) | Posted Speed (mph)
 | 25 | 30 | 35 | 40 | 45 | 50 | 55 | 60 | 65 | 70
10 | 105 | 150 | 205 | 270 | 450 | 500 | 550 | - | - | -
11 | 115 | 165 | 225 | 295 | 495 | 550 | 605 | 660 | - | -
12 | 125 | 180 | 245 | 320 | 540 | 600 | 660 | 720 | - | -

SIGN SPACING = X (1)
RURAL HIGHWAYS | 60 / 65 MPH | 800' ±
RURAL ROADS | 45 / 55 MPH | 500' ±
RURAL ROADS & URBAN ARTERIALS | 35 / 40 MPH | 350' ±
RURAL ROADS & URBAN ARTERIALS | 25 / 30 MPH | 200' ± (2)
RESIDENTIAL & BUSINESS DISTRICTS
URBAN STREETS | 25 MPH OR LESS | 100' ± (2)
(1) ALL SPACING MAY BE ADJUSTED TO ACCOMMODATE INTERCHANGE RAMPS, AT-GRADE INTERSECTIONS AND DRIVEWAYS.
(2) THIS SPACING MAY BE REDUCED IN URBAN AREAS TO FIT ROADWAY CONDITIONS.

CHANNELIZATION DEVICE SPACING (feet)
MPH | TAPER | TANGENT
50/70 | 40 | 80
35/45 | 30 | 60
25/30 | 20 | 40

BUFFER DATA
LONGITUDINAL BUFFER SPACE = B
SPEED (MPH) | 25 | 30 | 35 | 40 | 45 | 50 | 55 | 60 | 65 | 70
LENGTH (feet) | 155 | 200 | 250 | 305 | 360 | 425 | 495 | 570 | 645 | 730
TRANSPORTABLE ATTENUATOR ROLL AHEAD DISTANCE = R
HOST VEHICLE WEIGHT 9,900 TO 22,000 lbs.
HOST VEHICLE WEIGHT > 22,000 lbs.
< 45 MPH | 45-55 MPH | > 55 MPH | < 45 MPH | 45-55 MPH | > 55 MPH
100' | 123' | 172' | 74' | 100' | 150'
PROTECTIVE VEHICLE (WORK VEHICLE) = R
NO SPECIFIED DISTANCE REQUIRED

ROAD WORK AHEAD W20-1
ONE LANE ROAD AHEAD W20-4
BE PREPARED TO STOP W20-7B
W20-7A
BE PREPARED TO STOP W20-7B
W20-7A
CENTER LANE CLOSED AHEAD W20-5C
ROAD WORK AHEAD W20-1
DEVICE SPACING 1/2 DISTANCE FOR OPPOSING TRAFFIC DEVICES
WORK AREA
DEVICE SPACING 1/2 DISTANCE FOR OPPOSING TRAFFIC DEVICES
KEEP RIGHT R4-7B B/W
ROAD WORK AHEAD W20-1
W1-4(L)
BE PREPARED TO STOP W20-7B
W20-7A
W20-7A
BE PREPARED TO STOP W20-7B
ONE LANE ROAD AHEAD W20-4
ROAD WORK AHEAD W20-1

OPTIONAL IF 40 MPH OR LESS

NOTES
1. RECOMMEND EXTENDING DEVICE TAPER (L/3) ACROSS SHOULDER.
2. IF A SIGNAL IS PRESENT, IT SHALL BE SET TO "RED FLASH MODE" OR TURNED OFF DURING FLAGGING OPERATIONS.
3. FOR SPEED LIMIT OF 30 MPH OR LESS USE SIGN W1-3 IN LIEU OF SIGN W1-4.
4. MAINTAIN A MINIMUM OF ONE ACCESS POINT FOR EACH BUSINESS WITHIN WORK AREA LIMITS.
5. ALL SIGNS ARE BLACK ON ORANGE UNLESS OTHERWISE DESIGNATED.

LEGEND
FLAGGING STATION
TEMPORARY SIGN LOCATION
CHANNELIZING DEVICES
PROTECTIVE VEHICLE - RECOMMENDED
TEMPORARY SIGN LOCATION (5' MOUNTING HEIGHT)

INTERSECTION LANE CLOSURE - THREE LANE ROADWAY
NOT TO SCALE

FILE NAME: S:\Design R P& S\4-Standards\2-Plan Sheet Library\01-Published PSL\(TC) Work Zone Traffic Control\(TC-14) Intersection Lane Closure - Three Lane Roadway\TC-14.dgn
TIME: 6:59:33 AM
DATE: 1/3/2018
PLOTTED BY: liddelf
DESIGNED BY
ENTERED BY
CHECKED BY
PROJ. ENGR.
REGIONAL ADM.
REVISION | DATE | BY
REGION NO. | STATE: WASH
FED.AID PROJ.NO.
JOB NUMBER
CONTRACT NO. | LOCATION NO.
P.E. STAMP BOX | DATE
P.E. STAMP BOX | DATE
Washington State Department of Transportation
TRAFFIC CONTROL PLAN
Plot 1
PLAN REF NO. TC14
SHEET OF SHEETS

5.9. Tree Removal or Trimming

The types of tree work that trigger the requirement for a permit or approval vary by jurisdiction. Removal of trees is more commonly the trigger, sometimes with a size threshold such as only trees with trunks over four inches require specific approval. Even planting or pruning a tree in the ROW may require authorization or written permission in some jurisdictions.

However, before removing trees in the ROW, check with the local agencies for specific requirements. Project delays or fines fromz code compliance are more uncomfortable and usually require replacement of the tree with other saplings planted nearby, rather than just permission to remove a tree. Neighbors also often care deeply about trees in the ROW not on their properties and are likely to report illicit tree removal.

Some permitting agencies may issue a blanket permit to utility companies for tree trimming. For example, most power companies try to trim tree branches close to their power lines during good weather to avoid damage during bad and stormy weather. Since this is a routine maintenance operation, the AHJ may issue an annual permit to cover this operation.

5.10. Category III: Temporary or Long Term, Potential Disturbance of the Right-of-Way

This category includes any temporary or long-term construction activities within the public ROW. This type of activity has the potential to impact or disturb the ROW. However, often there is no disturbance within the ROW. It may include any of the following activities or a combination of:

- Hauling construction materials to or from the project site (i.e., hauling dirt, hauling demolished or hazardous materials)
- Carrying over the legal limit structures (i.e., manufactured homes)

These permits cover over-long, over-wide, over-weight, and construction vehicles over a specific traffic volume in an hour, day, or week, depending on the AHJ requirements. These permits tend to be relatively straightforward to obtain. The applicant should describe the vehicle's characteristics, the times, and dates for hauling, and show a map with a highlighted haul route where the vehicle can safely travel. Some trucking subcontractors prefer to get the haul permits. Keep in mind there is still a turnaround time for the approval of this type of permit and you should not wait until the last minute to apply.

Typically, a vehicle carrying an over the legal limit load must be accompanied by one or more pilot cars. A pilot car with safety lights must accompany the carrying vehicle if a structure (i.e., manufactured home) is over twelve feet wide, more than fourteen feet high, or over ninety feet long. In many states, if the width is over fourteen feet, two pilot cars are required, one leading and one following. The intent is to avoid the mishap shown in the photo on the next page.

OVERSIZE LOAD

5.11. How to apply for a Right-of-Way Permit?

We will discuss permits and the permitting process in more detail in Chapter 10. However, the ROW permits, and their processes are somehow different, so they are included in this chapter. In general, there are three ways, in no special order, the AHJ issues their ROW permits.

First, the ROW permit applications and documentation are typically generated in the Public Works Department, reviewed, and approved by its professional staff before the AHJ issues the permit.

Second, the ROW permit applications and documentation are generated as part of a larger residential or commercial project in the Public Works Department, reviewed, and approved by its professional staff before issuing the permit by the AHJ.

Third, the ROW permit applications and documentation are generated as part of a larger residential or commercial project in the Community and Development (AKA, Development Services) Department, reviewed, and approved by its professional staff before the AHJ issues the permit.

As mentioned earlier, when a ROW is part of a more considerable development (residential, commercial, or industrial), the development services or the community and development department, instead of the public works department, is usually the lead review and permitting agency. Regardless of the reviewing department, there should be close coordination between the public works and the community and development or development services departments for intaking a ROW application, reviewing and approving and issuing the permit, collecting fees, etc. We will discuss this subject further in Chapter 10.

Most jurisdictions have online applications available and may even have fully online submittal for permits. Start with finding the form for the type of ROW permit you seek. Remember that, per the preceding sections, there may be more than one category of ROW permits and, therefore, many different forms, so it is important to read thoroughly enough to ascertain you have chosen the correct form. If in doubt, it pays to call the permit center and ask questions. However, reviewing different applications and information before calling means the questions are more intelligent, and usually demand a thoughtful answer.

Fill out the form completely. Taking the time to fill it out right the first time can prevent misunderstandings and needless rounds of permit review. There may also be a cost associated with corrections by some agencies. Ensure to provide contact information and a descriptive enough narrative to describe the activities permitted under this permit fully.

Other supplementary submittal requirements may include a cost estimate to figure out the bond amount and, in some cases, permit fees depending on the AHJ. Also, payment of permit fees and proof of required business and contractors' licenses to pick up the permit.

Another additional submittal requirement for Category II mentioned above for ROW permits is a TCP, discussed in more detail in Section 5.8.

5.12. How long do ROW Permits Take?

Depending on the category of ROW permits, and the AHJ, the processing time from intake to issuance of a permit may vary from days to weeks. One mistake applicant often make is not budgeting enough time to acquire a ROW permit. Therefore, contacting the jurisdiction or checking their website is essential to see how long a review cycle will take. If possible, submit early to allow enough time for at least one round of comments without interfering with your construction schedule. Respond quickly to review comments without jeopardizing quality control to expedite the review. It is generally not helpful to check in more than once about a submittal because it will irritate and take more time from a reviewer who must balance your demands with multiple other developments.

5.13. After a ROW Permit is Submitted

After submission, plans are reviewed by at least one person, possibly multiple people, depending on the city or county and complexity of the project. Fire department representatives generally review any changes in relation to water design or fire hydrants. If the city has a dedicated traffic engineer, that person may review the TCPs. These reviewers prepare letters, memos, and/or redlined (marked up) PDF plan sets with any review comments.

Submittals that are not approved are returned to the developer to be reviewed, corrected, and resubmitted. It is important to address every comment made, even if you do not think it is correct or applicable to your project. Many jurisdictions recommend or require a comment response letter addressing those comments. This is your opportunity to explain how your plans meet the required standard. Submitting plans without correcting or explaining every issue invites a lengthy and costly review process.

Once all the reviewing organizations approve the permit, permit fees and bond amounts must be satisfied. The permit center may require evidence of the contractors' and business licenses before issuing the permit. Most permits have general and specific conditions associated with them. The conditions of the permit provide instructions to the contractor before, during, and after construction. The contractor should follow those instructions to avoid construction and schedule delays for closing the project and obtaining a certificate of occupancy. Make sure to review the conditions attached to the permit. It will explain to the applicant when and how to contract an inspector, whether a pre-construction meeting is required and who should attend, limits to working hours, and more.

CHAPTER 6: SINGLE FAMILY RESIDENTIAL AND ADU

6.1. Introduction

The previous five chapters discussed the information and data required to perform a sound site design and engineering plan preparation. This chapter discusses site design and engineering requirements for small projects, such as single family residential (SFR) and Accessory Dwelling Units (ADU). Although it is possible to have an SFR with a large building, extensive parking lots, swimming pools, and other facilities to trigger a large project construction site, the average SFR in the USA is about 2500 SF.

Therefore, this chapter focuses on the site improvements for smaller sites such as SFR, additions and ADU, and discusses the requirements and criteria for site improvements related to these smaller projects. In the following three chapters, we will discuss larger residential and commercial projects which do not qualify for small drainage requirements.

6.2. What is a Single Family Residential (SFR)?

A single-family residential (single-family detached) is a stand-alone structure on a lot intended for one family. Single-family residences differ from condominiums, townhomes, cooperatives, and multi-family homes. All are attached dwellings.

An SFR can be built on any legal lot. Generally, a legal lot is any whose dimensions and area are surveyed by a professional land surveyor registered in the state where the property is located. It is also officially recorded and approved by the county or other AHJ officials. Any professional surveyor should locate all corners and describe such a property in the field.

An SFR unit can be constructed on a residential approved lot within or outside a formal subdivision. As long as the lot is zoned residential and a physical address is assigned to the lot by the AHJ, the owner or their representative (i.e., applicant, contractor) should be able to obtain the necessary permit to construct an SFR.

6.3. What is an Accessory Dwelling Unit (ADU) or Mother-in-Law Unit?

An accessory dwelling unit can be added to an SFR either as an attached structure or detached. It depends on the size of the lot to ensure all required setbacks are met. ADUs may have additional restrictions that differ from one jurisdiction to another. Of course, the size of the lot to encounter setbacks and other AHJ requirements is critical in securing a permit. Other essential elements are potable water, a sanitary sewer, and utility availability. Also, the lot must have enough land to accommodate parking spaces for the SFR and ADU or, in some jurisdictions, be close to a public transit system such as public buses or light rail.

Be aware that even though the zoning and other AHJ requirements allow an ADU, the Homeowners Association (HOA) may not allow such construction. Suppose a residence is located within a subdivision with deed restrictions. In that case, the lot owner agrees at the time of acquiring the property that they abide by the deed restriction, which may not allow an ADU. Another example of such a deed restriction may be the number of stories and the structure height. Typically, the HOA Covenants and Restrictions overrule the AHJ in all those cases if their requirements are more stringent than the AHJ requirements.

6.4. Example 1 – SFR and ADU Drawing

The following drawing is an example site plan utilized by the City of Bellevue in King County, Washington State. These and other municipalities use modified versions of similar drawings to convey and guide their customers to the minimum requirements of their agency. The primary purpose of the example drawing is to address the SFR additions and ADU minimum requirements. However, many elements overlap and are required by other project types, such as short plat, subdivisions, and commercial projects.

The below sample drawing provides the minimum requirements for the AHJ to review. They serve excellent service for those who don't know much about how to secure a building permit. More importantly, they give a sample of what to present to the agencies to obtain a permit.

Let's go over some of the features in the following drawing. We will skip the simple ones and discuss some of the more challenging elements. We start with the front setbacks (#2 in the diagram). Since this SFR is located on the corner, both sides need to be set back from the street, which means there is less land for development. The rear-yard setback (#3) dimension is significant, perhaps due to the critical area element and required buffer. It is much larger than the side setback, since the backyard has a stream with a required buffer. The driveway (#9) width and details need to meet the AHJ standards and requirements that may differ for different agencies. Also, specific minimum requirements for the driveway distance (10) from the property line and the neighboring driveway could vary for various agencies. The critical area element

(#14), such as a stream, must be delineated by qualified professionals and surveyed so that the construction can be buffered away from the critical area as required by the AHJ.

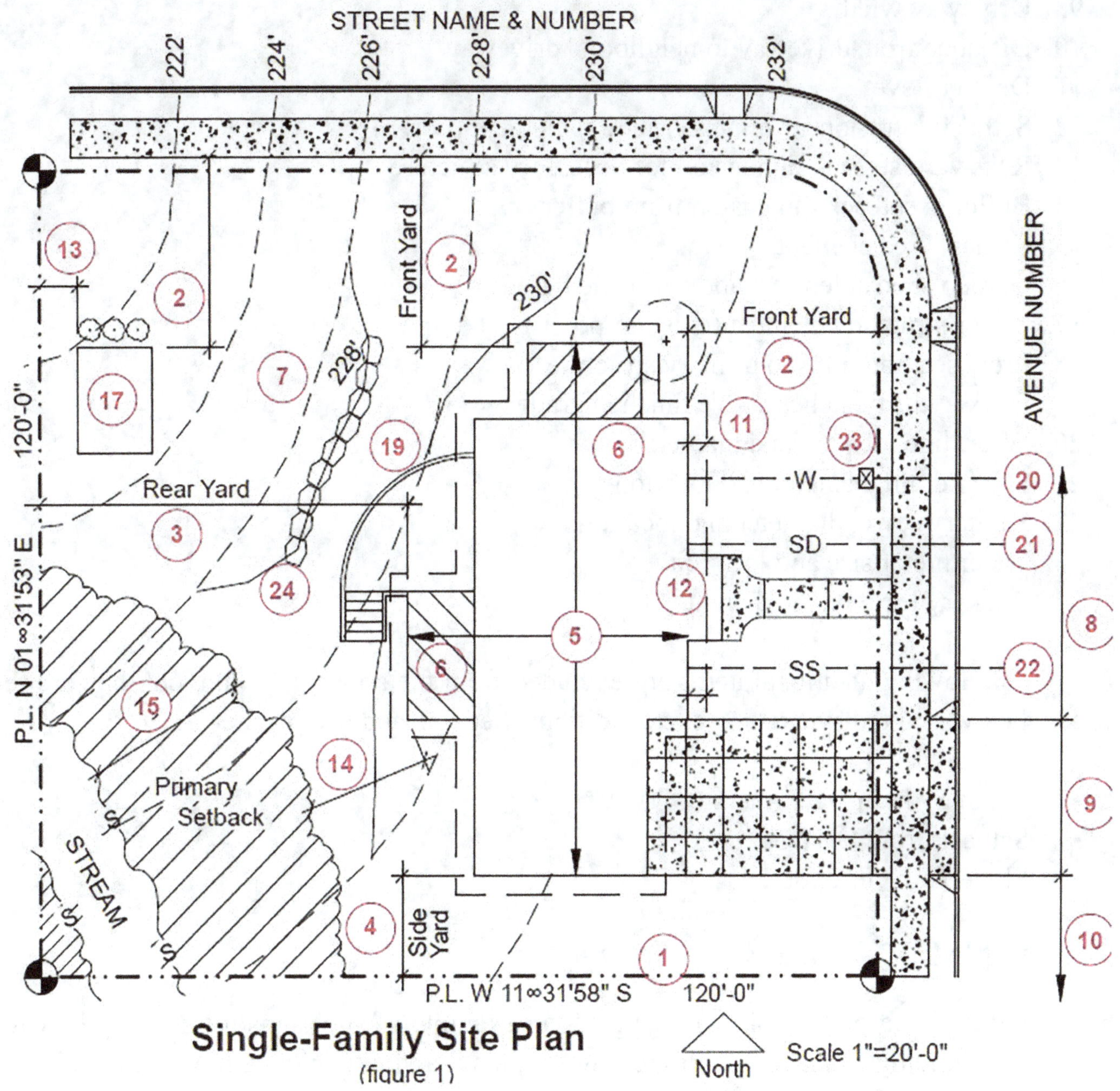

Single-Family Site Plan
(figure 1)

Numeric legend for key notes above

1. Property line
2. Front-yard setback dimension (applies to all structures)
3. Rear-yard setback dimension (for primary structure)
4. Side-yard setback dimension
5. Existing building footprint

6. Proposed additions or new structures
7. Existing and proposed contours (two ft. intervals)
8. Distance from driveway to face of curb/edge of paving
9. Driveway width
10. Distance from driveway to neighbors' driveway
11. Depth of eve
12. Setback intrusion by minor structural element (porch)
13. Rear-yard setback dimension for accessory structure
14. Building setback dimension from buffer
15. Critical area elements
16. Garden shed at ten feet above finished grade
17. Minor structural element (existing porch)
18. Proposed additions to primary structure
19. Deck at thirty inches above finished grade
20. Water line location and diameter
21. Storm drain diameter and location
22. Sanitary sewer diameter and location
23. Water meter size and location
24. Rockery location and height

Note: The following additional items are excluded from the above site plan, but they are required when warranted by existing site conditions. See the land use planner.

- Ordinary high-water mark (OHWM.)
- Setback from OHWM.
- Floodplain elevation

6.5. Example 2 – SFR and ADU Drawing

The second drawing on the following page is an example of site plans utilized by the City of Redmond in King County, Washington State. This drawing depicts a few additional proposed features in addition to the information shown on the example 1 plan. We discuss a few of these other features. The location of large trees to be protected is shown here, and the location of the temporary construction entrance that needs to be maintained during the construction. The area of the covered stockpile for construction materials and the location and setback dimensions for infiltration or dispersion trench. The additional elements shown in this drawing, the temporary erosion and sediment control (TESC), should be included in all construction plans.

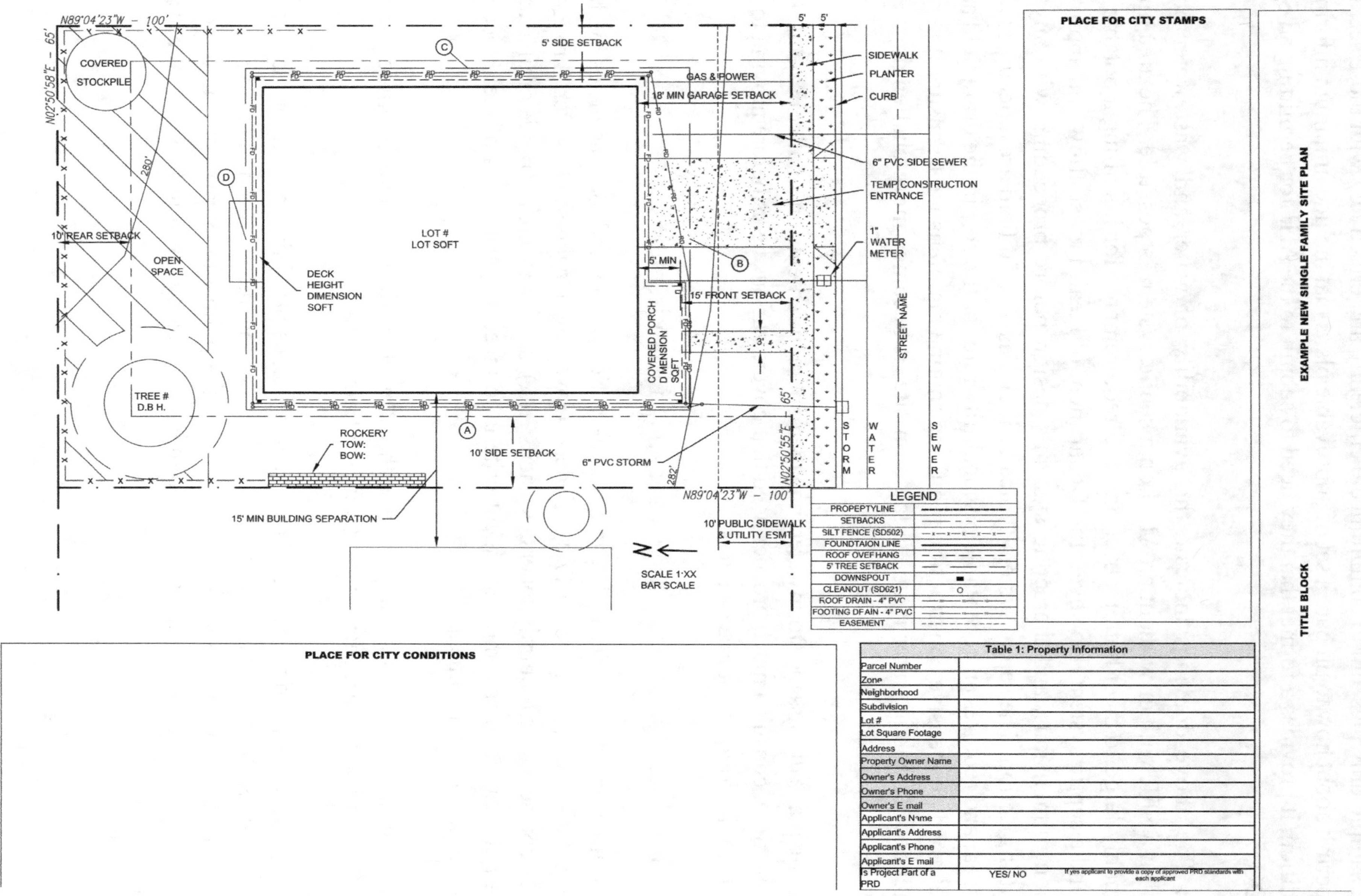

LEGEND	
PROPEPTYLINE	
SETBACKS	
SILT FENCE (SD502)	
FOUNDTAION LINE	
ROOF OVEFHANG	
5' TREE SETBACK	
DOWNSPOUT	
CLEANOUT (SD621)	O
ROOF DRAIN - 4" PVC	
FOOTING DFAIN - 4" PVC	
EASEMENT	

Table 1: Property Information		
Parcel Number		
Zone		
Neighborhood		
Subdivision		
Lot #		
Lot Square Footage		
Address		
Property Owner Name		
Owner's Address		
Owner's Phone		
Owner's E mail		
Applicant's Name		
Applicant's Address		
Applicant's Phone		
Applicant's E mail		
Is Project Part of a PRD	YES/ NO	If yes applicant to provide a copy of approved PRD standards with each applicant

These sample drawings aim to communicate efficiently and effectively with those seeking an SFR permit about how to prepare a set of approvable plans that include minimum information. In addition, it provides a format that they need to submit to the AHJ for permitting an SFR or an ADU.

Note: The above drawings indicate the availability of water and wastewater on site. Building a SFR with or without an ADU on a septic system is possible. The area must be reasonably large to meet both restrictions and requirements for installing a potable water well and a septic system. However, the Department of Health must allow and approve the location, dimensions, and the septic system's design before proceeding with any other permitting action. The septic system typically includes a drain field area, occupies a large land area, and requires an acceptable permeability rate for the soil in the drain field area. It is a relatively expensive construction requiring routine maintenance to ensure it performs as intended. If public water is unavailable on site, it becomes more challenging. Buyer, be aware!

These site plans discuss the stormwater management requirements on smaller projects. But, at the same time, consult a much broader need that covers various tasks before moving to larger projects in the following three chapters.

6.6. Stormwater Management System Design Requirements

The Clean Water Act of 1972 is delegated to the various states and local agencies by the Environmental Protection Agency (EPA), a Federal Government agency. However, through their elected representative (commissioners, councils, etc.), some agencies approve ordinances and accept policies that create more stringent criteria to meet the challenges of their specific regional needs and environment.

As we cover in more detail in the following three chapters, there are two design requirements related to stormwater management systems worth reviewing:

1) The water quality requirements. Water quality means providing natural or manufactured devices to filter the stormwater runoff before it infiltrates into the ground or discharges into a natural body of water. The intent is to prevent the polluted stormwater runoff from discharging into existing water bodies or wetlands and causing harm to fish and other aquatic habitats.

2) Water quantity or flow control requirements. Flow control means controlling either the peak flow and/or the runoff volume to avoid flooding downstream of the project. It may also mean properly designing the conveyor system to prevent downstream flow restrictions and therefore causing flow back up to the project site and the neighboring properties. The intent is to avoid or reduce the increased runoff peaks, volume, and loss of groundwater associated with the conversion of forested or grassland to impervious surfaces. The impacts are known to cause or contribute to downstream flooding, erosion, sedimentation, loss of aquatic habitat, loss of groundwater available for human consumption, and summer stream flows necessary for fish use.

Getting approval for SFR or ADU drainage requirements may seem simple. However, many permitting agencies have restricted rules and regulations to be followed by the applicant before accepting an application for an SFR construction permit. Some agencies allow concurrent submittal applications for site engineering to include stormwater management and building permits. Regardless, it is essential to understand those requirements and how to design and prepare construction plans to save time and money.

King County, Washington, probably has one of the most stringent criteria in the nation for the stormwater management system. We will discuss some criteria featured in the *King County Surface Water Design Manual*. The manual is about 1500 pages long and includes appendices A: Maintenance Requirements for Flow Control, Conveyance, and WQ Facilities; B: Master Drainage Plan Objective, Criteria, Components and Review Process; C: Simplified Drainage Requirements; and D: Construction Stormwater Pollution Prevention (CSWPP) Standards. The manual also includes details and specifications to cover almost any conceivable stormwater-related situation. The purpose of this chapter is not to teach the *King County Surface Water Design Manual* or even cover most items within it. The intent is to start this chapter with the *King County Surface Water Design Manual* and review some of its more common features that can be used anywhere else, with some modifications to fit a specific project. The reader is encouraged to use this manual as a reference or study it if they are more inclined to master stormwater management design and plan preparation.

Before discussing SFR and ADU example projects in this chapter, we cover some of the minimum requirements that trigger stormwater management system design, construction plan preparation, and permitting. Generally, if a project meets one or more of the following minimum criteria, regulatory agencies will require a drainage permit depending on the project's location and AHJ. The type of permit application and the technical information report (TIR) or drainage report could also differ depending on the two design requirements mentioned above.

The following criteria indicate requirements within the *King County Surface Water Design Manual*, and many cities within the county. Many other jurisdictions in other states, such as California, Florida, Oregon, and Texas, have similar measures to ensure that developments meet or exceed their environmental water quality and quantity requirements. It is imperative to contact the permitting agencies where the project is located for nuances of each permit requirement. However, knowing some of the criteria and design features may help the developers or design engineers be prepared to modify and adjust their design accordingly to meet AHJ requirements.

A. The project includes two thousand square feet of new impervious surface (concrete, asphalt, gravel, etc.), replaced impervious surface, or new plus replaced impervious surface.

Note: The replaced impervious surface is when the actual soil surface is exposed and disturbed. In other words, pavement repair or grind and overlay typically do not count toward the two thousand square feet. If an existing building caught on fire and the owner decided to remove the entire building, including the slab, all proposed areas need to be treated. If the owner chooses to build on top of the existing slab, no treatment is required, as approved by AHJ. Regardless, it is a good practice to connect the roof runoff to some form of infiltration system, if practical.

Also, some may ask why gravel is an impervious surface, since gravel surfaces can percolate water surface runoff. However, in environmental agencies' view, gravel surfaces can harden over the years, get compacted, dense, and serve as impenetrable surfaces.

B. The project includes seven thousand square feet or more of land-disturbing activities.

Note: This means that a lot with an area of five thousand square feet, but with the addition of two thousand square feet of offsite land disturbance, will meet this criterion. On the other hand, if the lot area is ten thousand square feet, but only five thousand square feet are being disturbed, it does not meet this criterion.

C. The project includes constructing or modifying a twelve-inch or larger diameter storm drain or ditch or receiving runoff from a twelve-inch or larger diameter drainage pipe or ditch.

Note: This criterion doesn't apply to an in-kind replacement of an existing twelve-inch or larger diameter pipe.

D. The project is adjacent to a flood hazard area, erosion hazard area, steep slope or landslide or seismic hazard area, critical drainage area, other sensitive areas, or related buffer-regulated area.

The following additional criteria are based on local agencies' and municipalities' requirements, and the values may vary between variance agencies.
E. The project includes earth cutting of more than four feet in vertical depth and earth filling of three feet in vertical depth.

F. The project includes fifty cubic yards or more of cut or fill.

Note: The amount of excavation can be different based on the project location and the area's topography. Regardless, the AHJ identifies the minimum amount.

G. The project involves the redevelopment of an existing high-use site for more than one hundred thousand dollars.

Note: Of course, this criterion directly relates to the locality and the condition of the economy. Therefore, the value can significantly differ based on the project location, the year, labor, and material cost.

In the following sections, we discuss some low impact design (LID) features under a simplified drainage design, since most single-family residences qualify for a simple drainage design.

6.7. What is Low Impact Design (LID)?

The United States Environmental Protection Agency (EPA) defines LID as:

> The term low impact development (LID) refers to systems and practices that use or mimic natural processes that result in the infiltration, evapotranspiration, or use of stormwater to protect water quality and associated aquatic habitat. EPA currently uses the term green infrastructure to refer to the management of wet weather flows that use these processes, and to refer to the patchwork of natural areas that provide habitat, flood protection, cleaner air, and cleaner water. At both the site and regional scale, LID/GI practices aim to preserve, restore, and create green space using soils, vegetation, and rainwater harvest techniques. LID is an approach to land development

(or re-development) that works with nature to manage stormwater as close to its source as possible. LID employs principles such as preserving and recreating natural landscape features, minimizing effective imperviousness to create functional and appealing site drainage that treat stormwater as a resource rather than a waste product. There are many practices that have been used to adhere to these principles such as bioretention facilities, rain gardens, vegetated rooftops, rain barrels and permeable pavements. By implementing LID principles and practices, water can be managed in a way that reduces the impact of built areas and promotes the natural movement of water within an ecosystem or watershed. Applied on a broad scale, LID can maintain or restore a watershed's hydrologic and ecological functions.

In other words, instead of collecting the stormwater runoff using conventional methods via storm sewer pipes and storage facilities such as ponds, and vault, LID uses a decentralized approach that disperses and/or infiltrates the runoff generated at the source within proximity to the original location.

The following sections cover more common LIDs typically used for smaller projects, mainly SFR sites. They are excerpts from the *King County Surface Water Design Manual* (KCSWDM) with some explanation from the author. However, the information retrieved from the KCSWDM in Washington State should apply to where the condition and requirements exist. As always, the AHJ should be contacted to ensure that the design meets the local jurisdiction standards and requirements.

There are numerous LID or stormwater BMP devices currently in use. We mention them in the following list in alphabetic order but provide example designs for three more commonly used ones shown highlighted below.

Bioretention
Bioswale or grassed swale
Cisterns
Drywells
Filter strip
Gravel filled or infiltration trenches
Ground surface depressions
Inlet pollution removal devices
Permeable or porous pavement
Rain garden

Rainwater harvesting
Rock pads
Sheet flow
Splash blocks
Tree box filters
Vegetated roofs (aka green roofs)

6.8. Drywells

Use of Drywells for Full Infiltration

Drywells are gravel-filled holes as opposed to trenches and, therefore, may allow for a more compact design for smaller sites and areas where the depth to the maximum seasonal high-water table is relatively deep (e.g., six feet or greater). The following Figure C.2.2.C illustrates the specifications for dry well infiltration systems as outlined below:

Note: This is a requirement for King County, WA, and other jurisdictions that adopted the KCSWDM. Other municipalities may have more stringent criteria depending on geography and geology.

1) When located in coarse sands and cobbles, dry wells must contain a volume of gravel equal to, or greater than sixty cubic feet per 1,000 square feet of impervious surface served.

2) In medium sands, drywells must contain at least ninety cubic feet of gravel per 1,000 square feet of impervious surface they serve.

3) Drywells must be at least forty-eight inches in diameter and deep enough to contain the gravel amounts specified above for the soil type and impervious surface they serve.

4) The gravel used for drywells must be 11/2-inch to three-inch washed drain rock. The drain rock may be covered with backfill material as shown in the following Figure C.2.2.C or remain exposed at least six inches below the ground surface.

5) Filter fabric or geotextile must be placed on top of the drain rock (if proposed to be covered with backfill material) and on the drywell sides before filling with the drain rock.

6) Spacing between drywells shall be a minimum of ten feet.

7) Drywells must be setback at least fifteen feet from buildings with crawl space or basement elevations below the overflow point of the dry well.

Drywell Infiltration System - Example

Let's say for an 800 SF addition such as ADU,
Drywell dia. = 4 feet
Surface area = (Pi) x R x R = 3.14 x 4 = 12.56 SF

In coarse sands and cobbles:

(60 CF / 1000 SF) x 800 SF = 48 CF.
Minimum drywell depth = 48 CF / 12.56 SF = 3.82 feet
Let's use H = 4 feet
Proposed volume = 12.56 x 4' = 50.24 CF > 48 CF O.K.

In medium sand:

(90 CF / 1000 SF) x 800 SF = 72 CF.
Proposed depth = 72 CF / 12.56 SF = 5.73 feet
Let's use H = 6 feet
Proposed volume = 12.56 x 6' = 75.36 CF > 72 CF O.K.

FIGURE C.2.2.C TYPICAL DRYWELL INFILTRATION SYSTEM

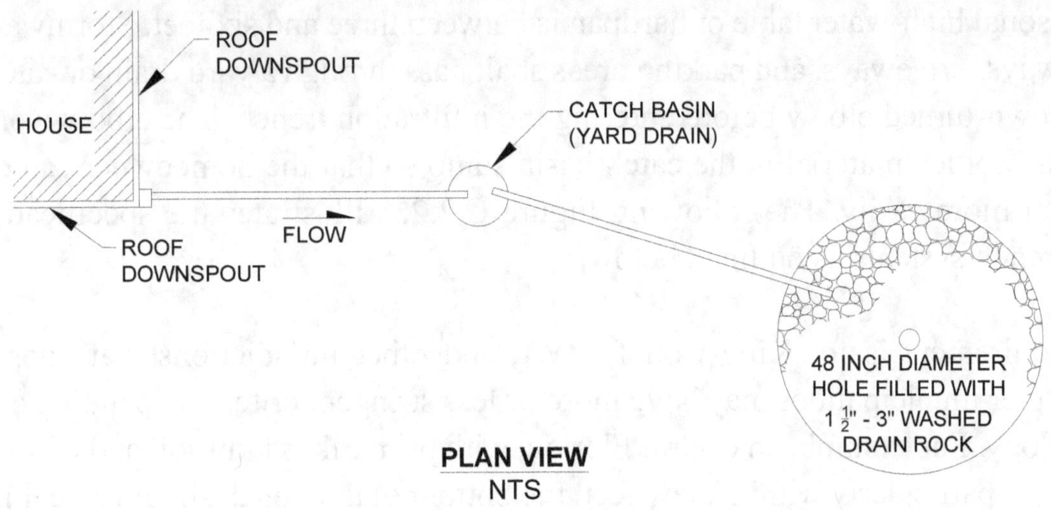

PLAN VIEW
NTS

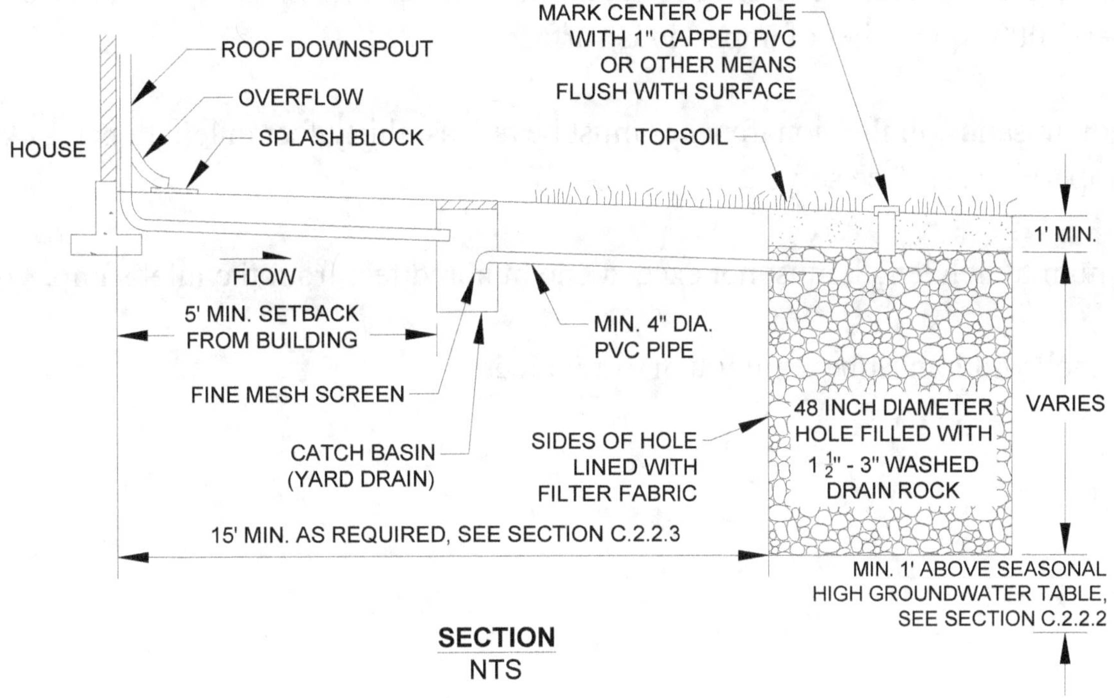

SECTION
NTS

6.9. Gravel Filled or Infiltration Trenches

Use of Gravel-filled Trenches for Full Infiltration

Gravel-filled trenches (AKA "infiltration trenches") are a good option where the depth to the maximum seasonal high-water table or hardpan is between three and six feet. Stormwater run-off from roadways, driveways, and parking areas shall pass through a yard drain or catch basin fitted with a down-turned elbow before entering the infiltration trench. The elbow works as a skimmer to trap spilled material in the catch basin sump so that the homeowner can cleanup spilled material more easily. The following Figure C.2.2.A illustrates the specifications for gravel-filled trench systems as outlined below:

Note: This requirement is for King County, WA, and other jurisdictions that adopted the KCSWDM. Other municipalities may have more or less stringent criteria depending on geography and geology. For example, in central Florida, with primarily sandy soil and a low probability of hardpan, particularly within a few feet, the bottom of the trench should be a minimum of two feet above the SHWT.

1. When located in coarse sands or cobbles, infiltration trenches must be at least twenty feet in length per 1,000 square feet of impervious surface served.

2. In medium sands, infiltration trenches must be at least thirty feet in length per 1,000 square feet of impervious surface served.

3. Maximum trench length must not exceed one hundred feet from the inlet sump.

4. The trench width must be a minimum of two feet.

5. Fill the trench with at least eighteen inches of 3/4-inch to 11/2-inch washed drain rock. The drain rock may be covered with backfill material as shown in Figure C.2.2.A or remain exposed at least six inches below the lowest surrounding ground surface, as shown in Figure C.2.2.B.

6. Filter fabric or geotextile must be placed on top of the drain rock (if proposed to be covered with backfill material) and on the trench sides before filling with the drain rock.

7. Spacing between trench centerlines must be at least six feet.

8. Infiltration trenches must be setback at least fifteen feet from buildings with crawl space or basement elevations below the infiltration system's overflow point.

Gravel-filled Trench for Infiltration System - Example

Let's say for an 800 SF addition such as ADU,

In coarse sand or corbel:

Required trench length (L) = (20 ft. / 1,000 SF) x 800 SF = 16 ft.

Use twenty-feet infiltration trench per Figure 6.2.2.A. or 6.2.2.B. > 16 ft.   O.K.

In medium sand:

Required trench length (L) = (30 ft. / 1,000 SF) x 800 SF = 24 ft.

Use 30 feet infiltration trench per Figure 6.2.2.A. or 6.2.2.B. > 24 ft.   O.K.

FIGURE C.2.2.A TYPICAL TRENCH INFILTRATION SYSTEM

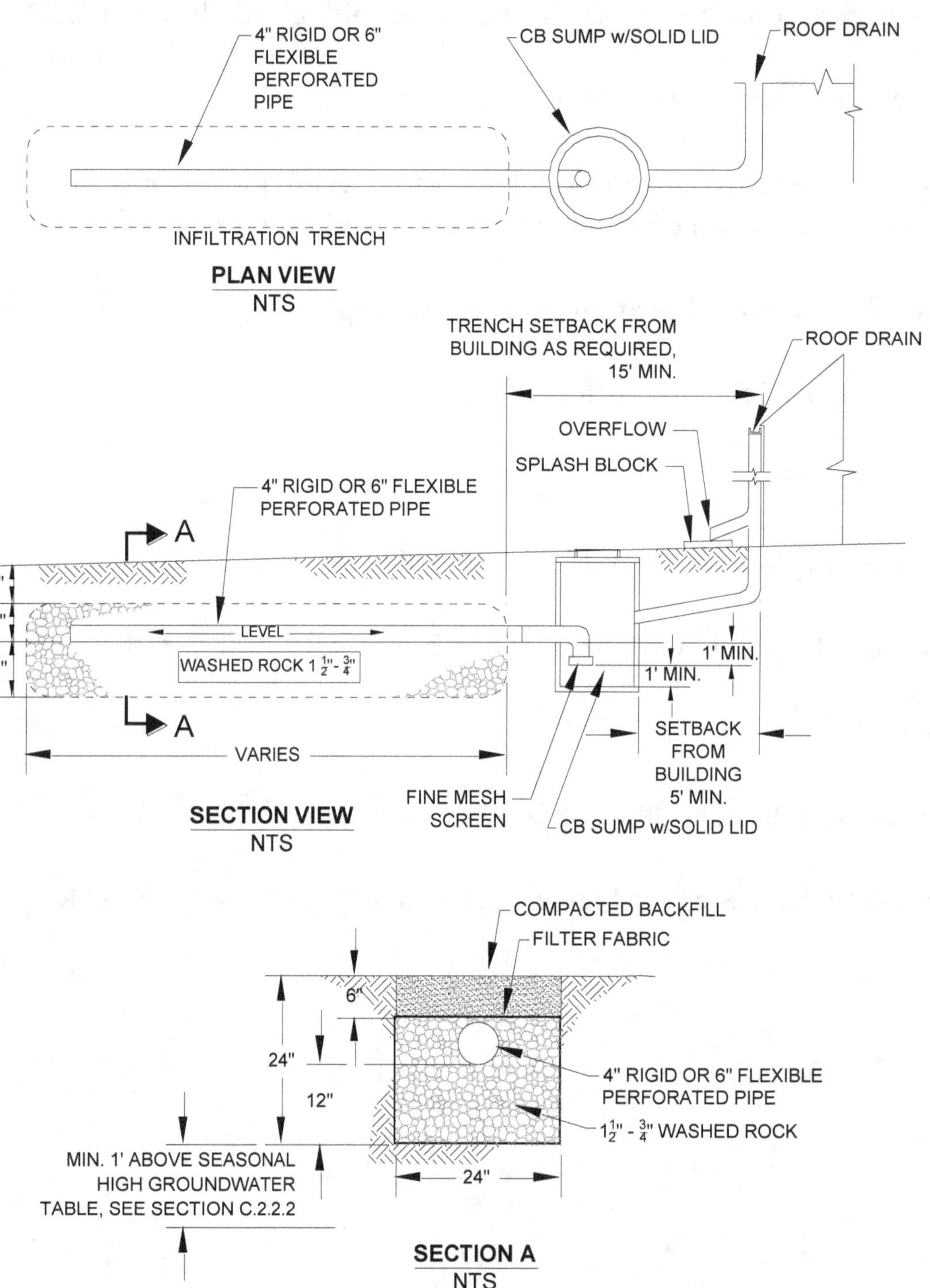

FIGURE C.2.2.B ALTERNATIVE TRENCH INFILTRATION SYSTEM

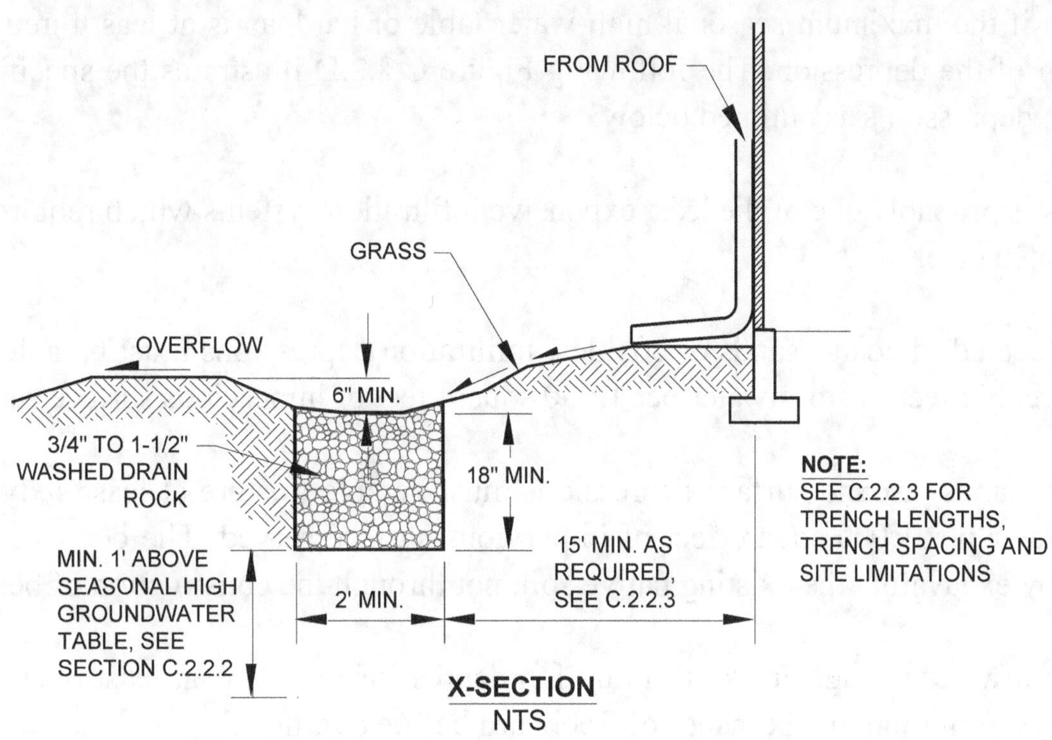

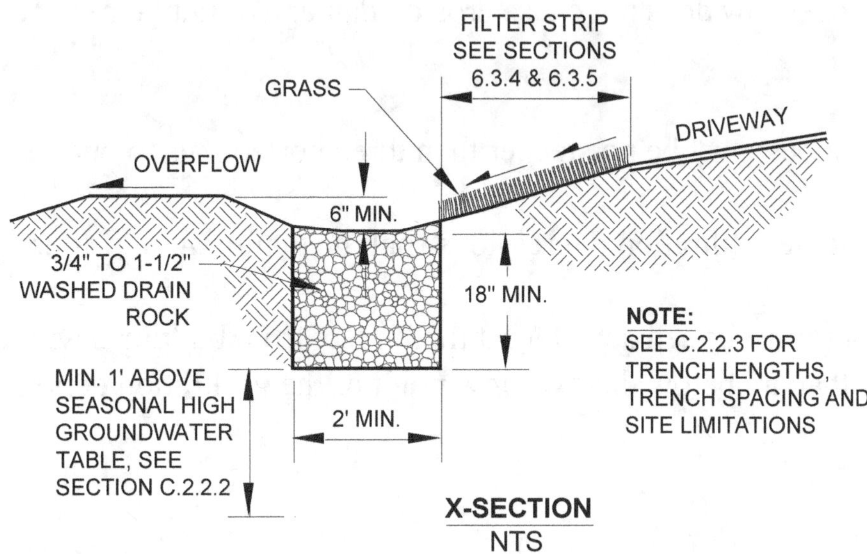

6.10. Ground Surface Depressions

Use of Ground Surface Depressions for Full Infiltration

Ground surface depressions (also called "infiltration depressions") are another option for full infiltration if the maximum seasonal high-water table or hardpan is at least three feet below the bottom of the depression. The following Figure C.2.2.D illustrates the specifications for infiltration depressions as outlined below:

Note: This is probably one of the least expensive infiltration systems which require minimum or no maintenance.

1) When located in coarse sands or cobbles, infiltration depressions must be able to store at least forty cubic feet of stormwater per 1,000 square feet of impervious surface served.

In medium sands, ground surface depressions must be able to store at least sixty cubic feet of stormwater per 1,000 square feet of impervious surface served. The depression must be achieved by excavating the existing native soil, not through the construction of berms.

2) The stormwater storage areas of infiltration depressions must be at least twelve inches in depth, with a minimum of six inches of freeboard before overflow.

3) The depression overflow point must be at least six inches below any adjacent pavement area and be situated, so overflow does not cause erosion damage or unplanned inundation and cause flooding.

4) The depression side slopes must be no steeper than three horizontals to one vertical.

5) Spacing between multiple infiltration depressions shall be a minimum of four feet.

6) Infiltration depressions must be setback at least fifteen feet from buildings with crawl space or basement elevations that are below the overflow point of the infiltration depression.

7) Infiltration depressions may be any size or shape, provided the above specifications and the minimum requirements are met.

8. The ground surface of the infiltration depression must be vegetated with grass or dense ground covers.

Ground Surface Depression - Example

Let's say for an 800 SF addition such as ADU,
Depth of the depression = 2 feet

In coarse sands or cobbles:

(40 CF / 1000 SF) x 800 SF = 48 CF.
Minimum depression area = 48 CF / 2 feet = 24 SF

Let's use a 4 feet x 6 feet rectangular area for the top.
The bottom area will be larger depending on the side's slopes,

Proposed volume > 24 SF x 2 feet = 48 CF O.K.

In medium sand:

(60 CF / 1000 SF) x 800 SF = 72 CF.
Minimum depression area = 72 CF / 2 feet = 36 SF

Let's use a 6 feet x 6 feet square area for the top.
The bottom area will be larger depending on the side's slopes,

Proposed volume > 36 SF x 2 feet = 72 CF O.K.

FIGURE C.2.2.D TYPICAL GROUND SURFACE DEPRESSION INFILTRATION SYSTEM

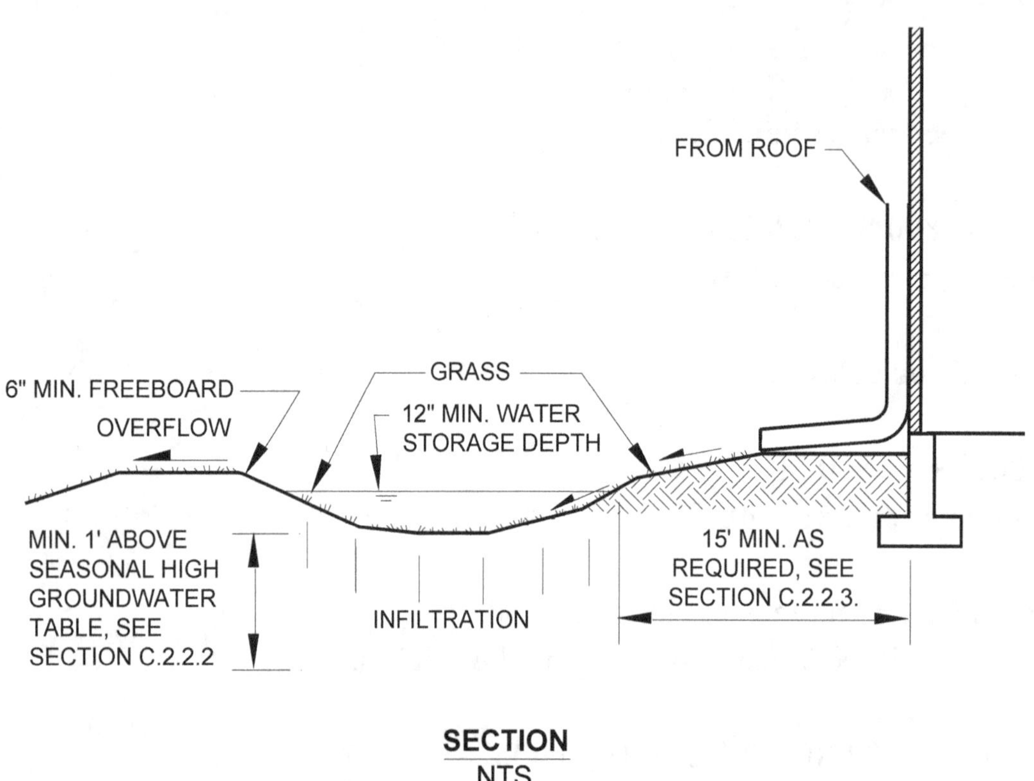

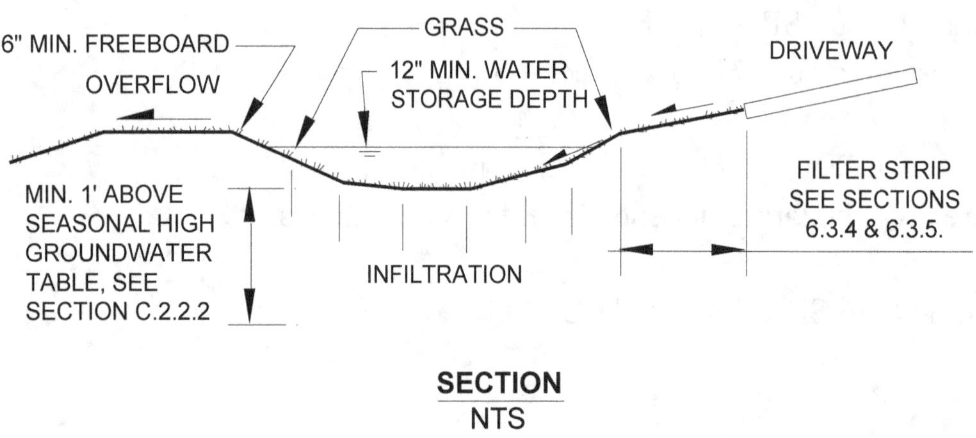

6.11. Inspection and Maintenance Requirements

The infiltration devices are designed to infiltrate the stormwater management flow control BMP to mitigate the stormwater quality and quantity impacts of some or all of the impervious surface areas of the project. For the infiltration device to function as it was intended and designed, they need to be appropriately constructed per the design detail and specs. They also must be maintained on a routine basis to extend the device's longevity. Since infiltration is a method of soaking runoff from an impervious area into the ground, the soil condition around the infiltration device must be reliably able to absorb water into the soil for a reasonable number of years.

The following minimum maintenance procedure must be performed to guarantee the life expectancy of the infiltration system. As always, the operator should follow the AHJ rules and regulations as they apply.

1) Inspect the infiltration devices annually and after major storm events to identify and repair any physical defects.

2) Maintenance and operation of the system should focus on ensuring the system's viability by preventing sediment-laden flows from entering the device. Excessive sedimentation will result in a plugged or non-functioning facility.

3) Keep the areas that drain to infiltration devices well swept and clean to enhance the longevity of these devices. For roofs, frequent cleaning of gutters will reduce sediment loads to these devices.

4) Remove sediment accumulation annually or more frequently, if necessary, particularly if an inlet is associated with the infiltration device. Prolonged ponding around or atop a device may indicate a plugged facility.

5) Replace the infiltration device if it becomes plugged and non-performing. Preferably, retest and redesign to ensure the new system can still perform as an infiltration device.

CHAPTER 7: SMALL RESIDENTIAL SUBDIVISION

Project 2: Magnolia Village

7.1. Introduction

This chapter discusses the design and construction plans prepared for a small sixteen-lot residential subdivision. Project 2, featured in this chapter, starts from vacant land to the survey map to a final completed construction plan, the only example in this book that provides comprehensive design and construction plans for the entire project. Shasti Enterprises LLC purchased this property in 2005 when the economy was at its peak performance. The construction industry was booming, and most residential and commercial properties were bought and sold at their highest price after being on the market for a short period. Of course, that depended on the location and proximity to public and private amenities.

This 4.78-acre property located in Central Florida is not relatively high but dry. It is not in a hundred-year flood zone, and there are no wetlands on the property. The site was zoned SFR. The biggest flaw of this property was that it was landlocked, and there was no available legal access to the property. This situation is not normal and fortunately does not happen often. Most regulatory agencies such as towns, cities, and counties in charge of permitting are mindful of this type of situation and do their best to ensure that this mishap does not occur. Regardless, this property was landlocked. The only way for the buyer to purchase the property was to perform full-blown due diligence. It was imperative to find ways to obtain access to the property without going thru a lengthy and time-consuming process or an expensive legal battle.

7.2. Site Data

Here is a summary of the site data:

•	Location	Central Florida
•	Dimensions	314.55 feet Wide x 673.20 feet Long
•	Area	4.78 acre (+/-)
•	FEMA	Flood Zone "C" Community Panel No. 120113 Panel 0005-B, Revised date: 4-29-1983

• Zoned	R1 - A (Single Family Residential)
• Access to public street	None
• Topography	Fairly flat
• Critical areas	No wetlands or other water bodies such as lake, stream, or pond
• Trees	Small to medium-size trees and vegetation
• (No majestic oak or other significant trees)	
• Potable water	Available within a short distance
• Sanitary sewer	Available within a short distance

7.3. R1-A (Single Family Residential)

Every jurisdiction has its zoning codes, which establish minimum requirements for laying out a site. Therefore, one of the early criteria to consider should be the zoning district where the project is situated and the minimum requirements and restrictions for that specific zoning district.

The following is a summary of the minimum requirements for zone R1-A, where this project was located:

• Minimum lot size	7,500 SF
• Minimum ROW width	50 feet
• Minimum lot width	75 feet
• Minimum building setbacks	25 feet front
• 30 feet rear	
• 10 feet sides	

Based on the above-stated data obtained from the various resources, including AHJ, the following sections and corresponding figures explain how to prepare preliminary lot layout computations and lead toward final site construction plans.

7.4. Stormwater Design Parameters

The most challenging part of designing and preparing a site engineering and construction plan is perhaps related to the stormwater management system. This discipline consumes most property and requires meeting or exceeding multiple agencies' rules and regulations. That is why it is crucial for the engineer and the designer to understand the requirements related to the property before diving deep into the design.

For this project, the following stormwater management design criteria defined:

1)   Water Quality Treatment (Dry Detention Pond)

We designed this project as a dry pond and set the pond bottom a minimum of two feet above the SHWE. The requirement was to provide online storage treatment volume based on a half-inch rainfall over the project area.

2)   Open vs. Closed Drainage Basin

The Southwest Florida Management District (SWFWMD) defines a closed drainage basin as "a watershed in which the runoff does not have a surface outfall up to and including the 100year flood level". This site is within SWFWMD in an open drainage basin, meaning the project has a positive outfall. The stormwater runoff discharges from the site much sooner before it reaches the hundred-year flood level.

3)   Water Quantity Requirement

The requirement is to meet or exceed the twenty-five-year, twenty-four-hour peak discharge, since the project is within an open drainage basin. In other words, we need to ensure that storm runoff release from the site during post-development is equal or preferably below the pre-development peak discharge for a twenty-five-year, twenty-four-hour storm event for our specific location.

Note: It is well worth mentioning that there are many different ways to design a site engineering project. The stormwater management system designed for this project used the conventional method of collecting the stormwater runoff via inlets and pipes system and storing it in a dry detention pond. The proposed pond bottom, side slopes, and infiltration system have provided water quality treatment.

One can look at designing either partially or the complete project using LID. For example, let's assume each lot provided by an infiltration system percolated its stormwater runoff from the roof and pavement area to the ground. We can also design the sidewalk, driveways, and roadway as porous pavement. This design is entirely acceptable and doable with the type of soil and topography of this project. It may even increase the number of lots by two, from sixteen to eighteen lots. The downside is that it will require much more monitoring and maintenance requirements. In this case, the AHJ accepted to take over the responsibility of the roadway and ponds. They may not agree to accept the responsibility of a porous pavement roadway as a public roadway. Regardless, we should explore various design options to reach an optimized design to serve the client and the public.

7.5. Preliminary Lot Computation

We divide the property length by the minimum lot width, which results in 670 feet / 75 feet, which equals 8.993 lots. This number represents the maximum number of lots that can fit along one side.

We consider dry ponds since this property is in a well-drained dry land and has a reasonable permeability rate per the geotechnical report. Therefore, we allocate a portion of the property for dry ponds.

Let's assume that the 0.993 reminders from the lot width will be used for the proposed dry pond along each side.

The approximate number of lots is eight lots per side, multiplied by two rows of lots that fit per the minimum lot dimensions, resulting in sixteen lots total.

Dividing the property width by two results in 313 feet / 2 = 156.5 feet. This number includes a minimum of twenty-five feet width for the proposed ROW.

Therefore, maximum property width = 156.5 feet − 25 feet = 131.5 feet

Let's say 130 feet

That results in a proposed lot area of 130 feet x 75 feet = 9,750 SF > 7,500 SF (the minimum lot area) O.K.

7.6. Preliminary Ponds Computation

Water Quality Requirement

The minimum volume required to meet the water quality requirement using a dry pond for detention = total project area x 0.5" = 209,959 SF (conservative) x 0.5" x 1'/12" = 8,748 CF. Typically, the project area is less than the total site area.

Assuming a shallow, one-foot-deep pond, the pond's area should be at least 8,748 SF.

Assuming two interconnected ponds that are located at the low area on both sides of the entrance,

Each pond width is 8,748 SF / (313 feet – 50 feet) minimum, which equals 33.3 feet.

The actual pond width will be longer than 33.3 feet since we will make the entrance a few feet wider than 50 feet to accommodate a landscaping island, sign, etc.

Check: 8 x 75 feet + 33.3 feet = 633.3 feet < 670 feet (total property length) O.K.

7.7. Preliminary Lot Layout

As explained in Chapter 2, a surveying map is one of the first steps in preparing a preliminary site plan. Figure 7-1 on the next page is a survey map prepared by a licensed land surveyor indicating boundary, topographic, and trees survey for this site. It also includes two-foot contours. Only four contour lines are shown for this site since it is a relatively flat site.

Note: We typically provide the following disclaimer note on the survey sheet when supplied by anyone other than us, the EOR. The owner/developer or the contractor provides the survey information. The EOR is not responsible for the accuracy of the provided information.

Based on the above computations and the minimum requirements, conceptually, a lot layout can be drafted, as shown in Figure 7-2 on the following page after Figure 7-1.

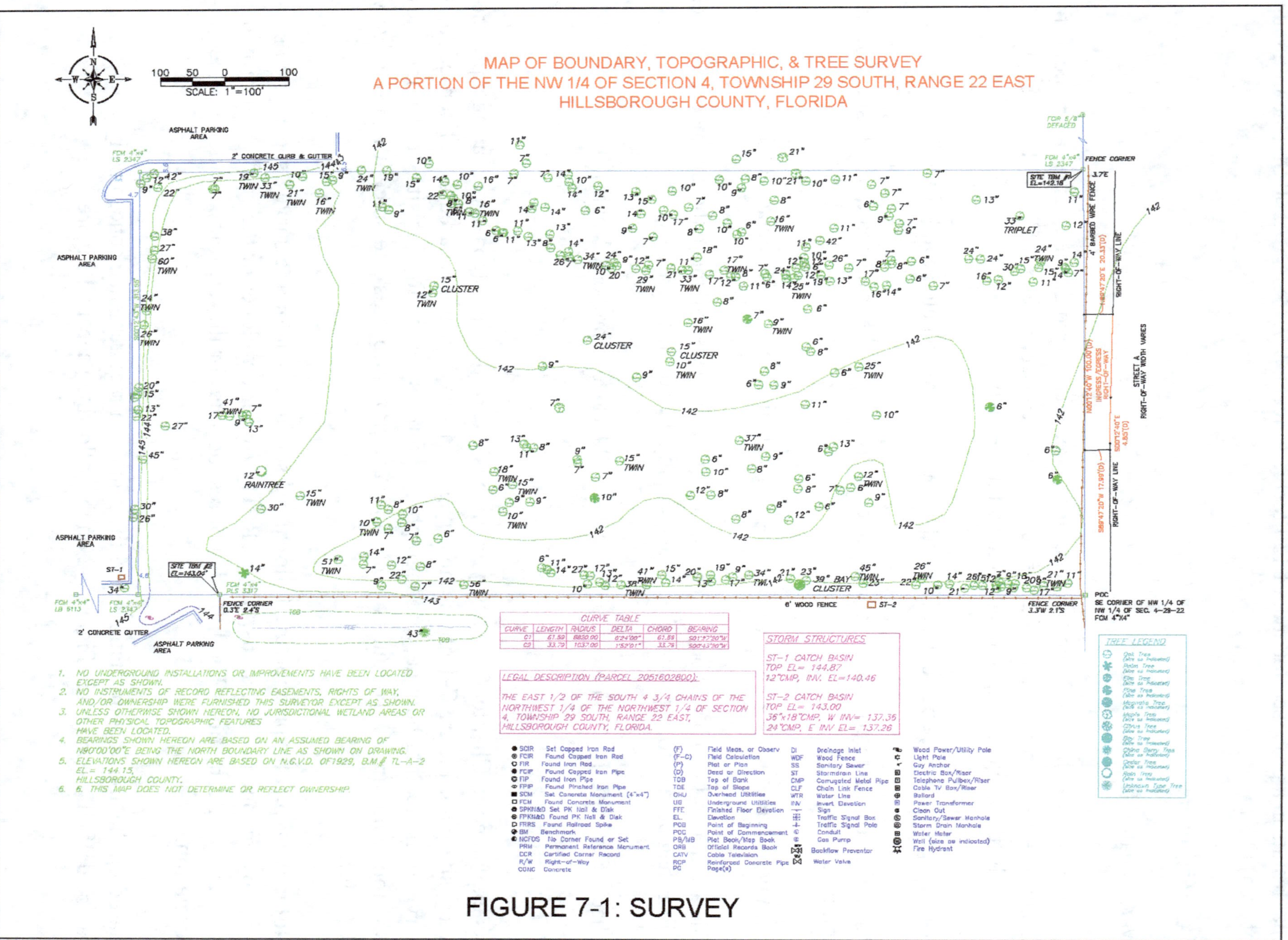

FIGURE 7-1: SURVEY

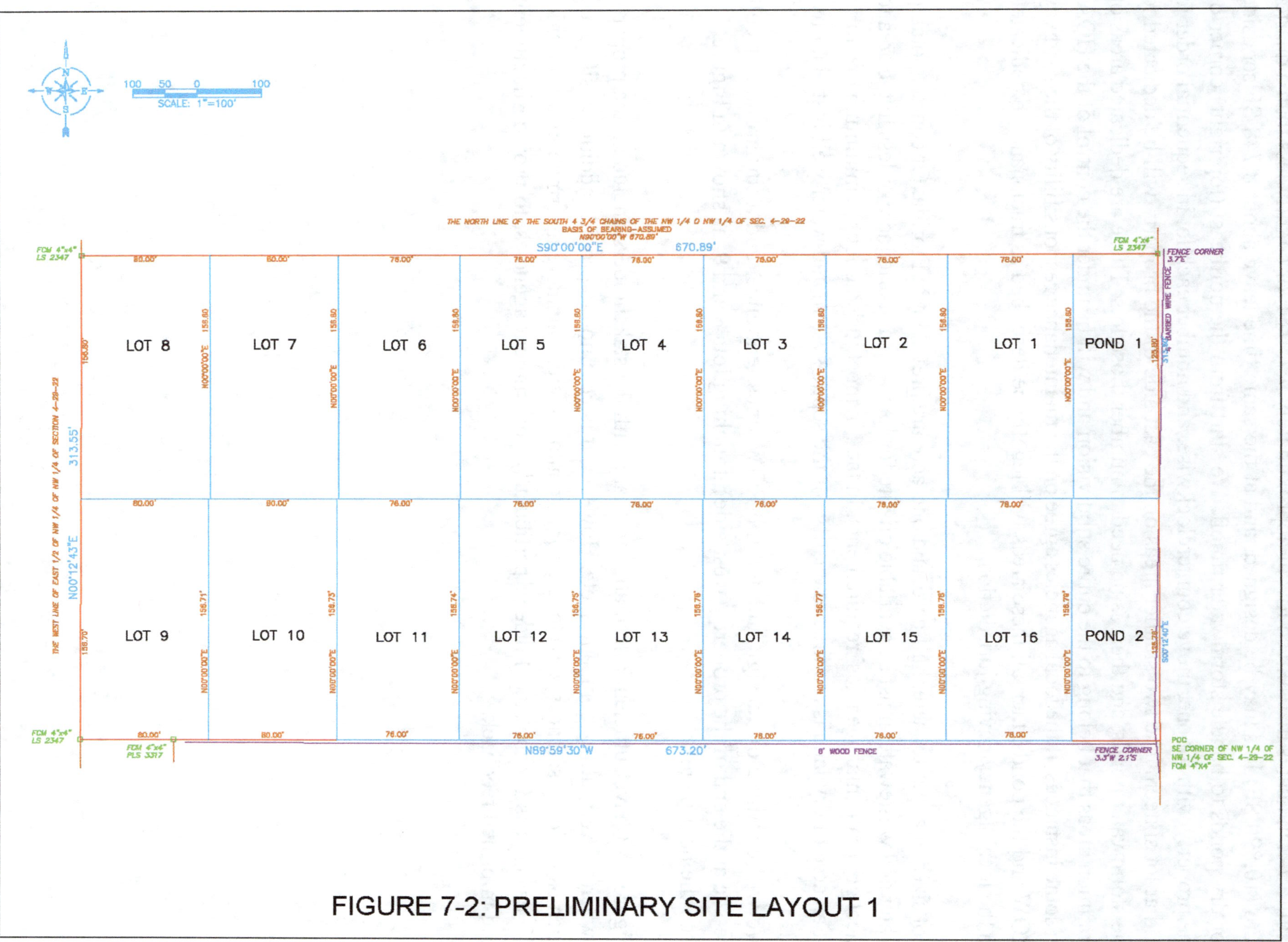

FIGURE 7-2: PRELIMINARY SITE LAYOUT 1

Figure 7-2 indicates that sixteen lots of approximately 9,750 SF or slightly less than a quarter acre (10,890 SF) lots may be designed, and at the same time leave about 8,748 SF set aside for dry ponds to handle stormwater runoff. We start from the northeast (top right) corner of the property with lot one, move counterclockwise, and number each lot in sequential order to end at lot sixteen. The two smaller areas on the east side of the property will be dedicated to the stormwater ponds, and there is no need to number those lots. These are dedicated areas or common areas that belong to the entire subdivision and will be under the control of the HOA. In some instances, if the local agencies agree on the maintenance responsibility of the roadway ROW and the ponds, then all these areas will legally be described as tracks and be dedicated to the local agency having jurisdiction.

Lot sizes of a quarter acre, half-acre, and one-acre are prevalent in Florida. Of course, potable water and wastewater must be available on-site for lots with a quarter acre or less in the area to meet governmental agencies' requirements. In fact, in most jurisdictions, a minimum of a third of an acre of dry land is required to qualify for the septic system and water well installations.

We will continue our design and add the ROW lines, entrance, and a turnaround or cul-de-sac at the end of the property (west side), touching lots 8 and 9, as shown in Figure 7-3 (see the next page).

Figures 7-3 have sufficient information to share with the stakeholders, including the property owner(s), AHJ, utility companies, etc. However, it is common to provide additional site data, such as building setbacks, envelopes, ponds contours, etc. The more information we provide, the easier it is for the stakeholders, particularly government agencies, to provide intelligent and informed feedback.

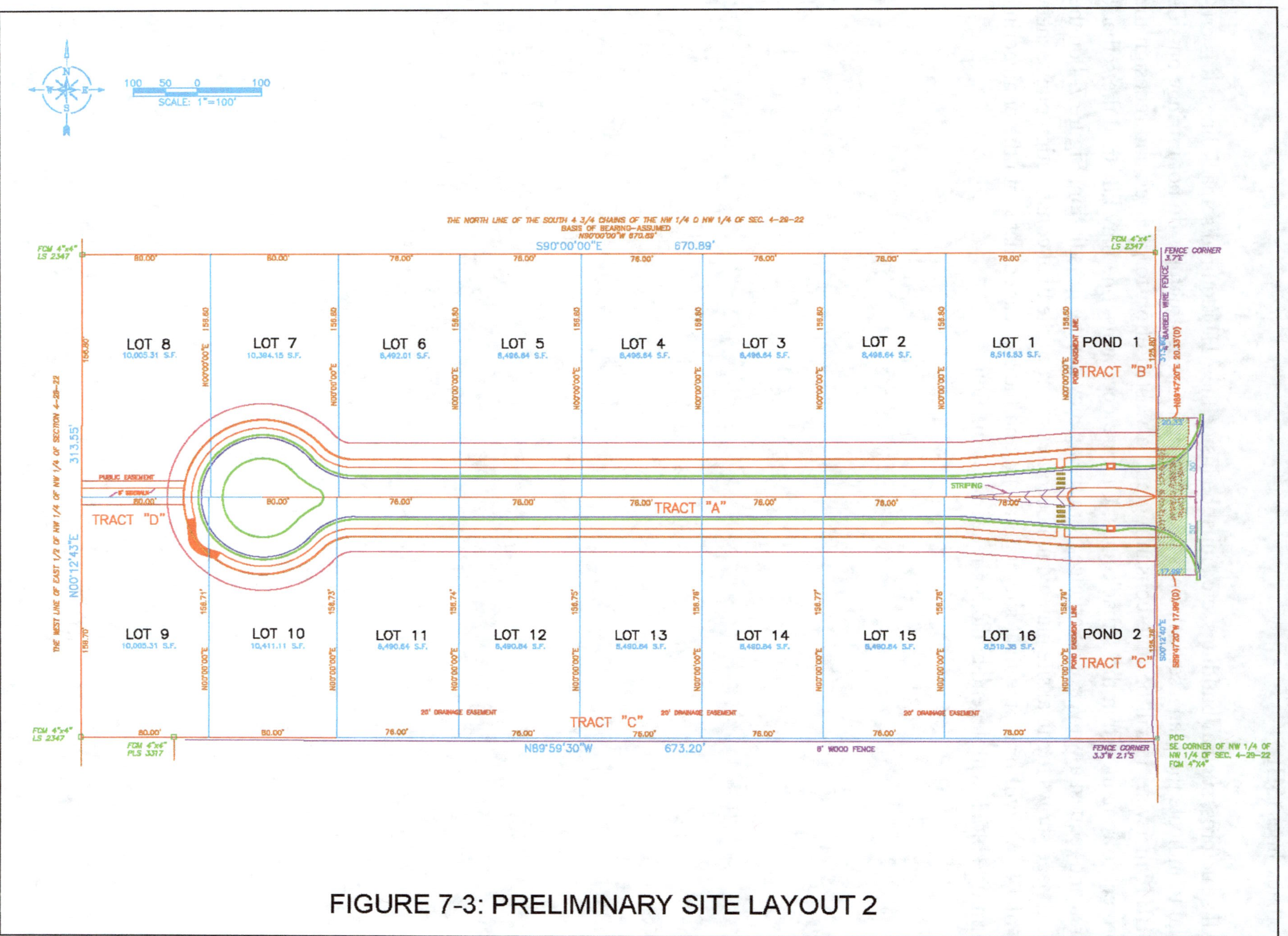

FIGURE 7-3: PRELIMINARY SITE LAYOUT 2

Setback lines (SBL)

All structures, such as buildings, etc., must be inside the AHJ required setback lines. It is crucial to indicate the plan's setback lines to ensure permanent structures do not infringe outside SBL. Otherwise, it will be time-consuming and a wasted effort to go back and redesign the project to remain within the setback lines. On rare occasions, one may advance the setback area and request a waiver or a variance from the agency having jurisdiction. Figure 7-4 on the next page shows the final site plan, which becomes a part of the final construction plans, sheet 3 of 14. We will bring the final site plan layout and the associated data into the engineering firm title block with legend and other related information.

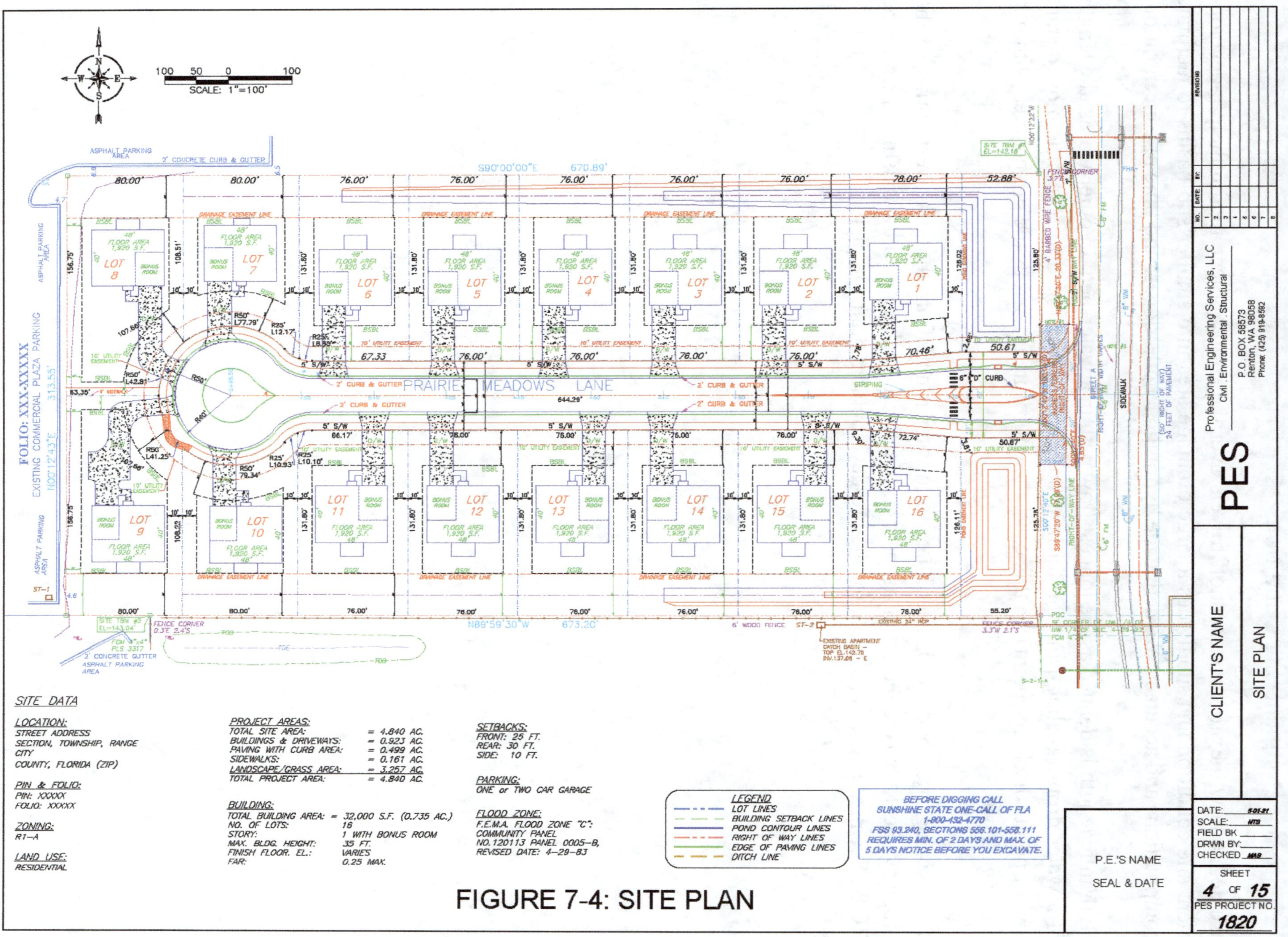

FIGURE 7-4: SITE PLAN

7.8. Pre-development Site Conditions

We define how the stormwater runoff travels over the entire project area in the pre-development site condition and subdivides a large drainage area into smaller sub-basins to facilitate computations and accuracy. We define the hydraulic length, slopes, and time of concentration.

Hydraulic length is defined as the longest flow path in a watershed. In other words, the length of a water drop traveling from the furthest point in the watershed to reach the most distant lowest point (discharge point). In this case, the entire site is used as the project area with one drainage basin.

The existing site condition is such that stormwater runoff from the northwest corner of the property sheet flows to the southeast corner. Finally, it discharges to the existing ditch in the ROW. See next page, Figure 7-5, for the pre-development drainage plan.

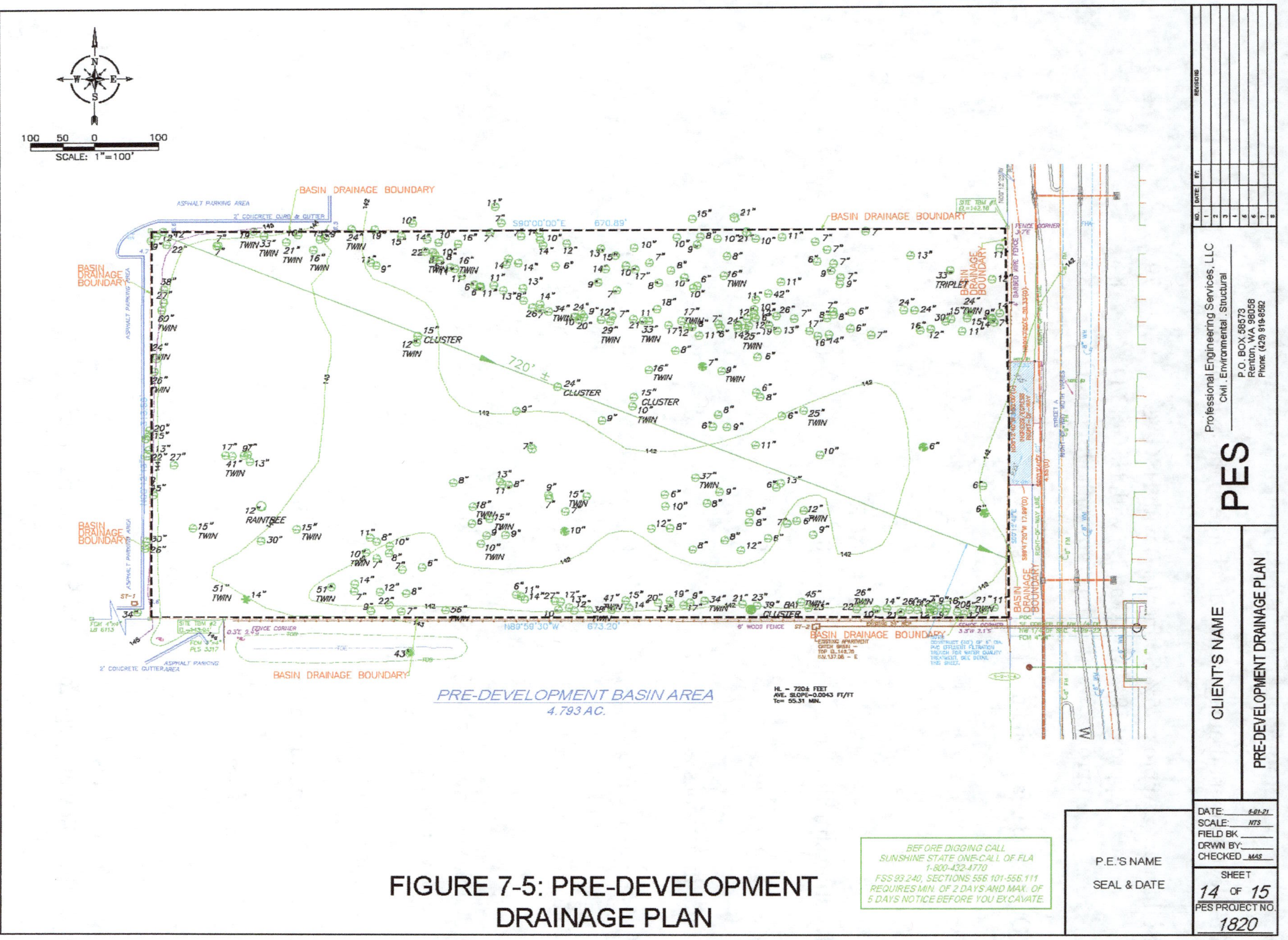

FIGURE 7-5: PRE-DEVELOPMENT DRAINAGE PLAN

7.9. Pre-development Data and Computations

Pre-development drainage area: (discharge to the existing low area)

		CN
Soil type:	Approx. 25% Seffner (47) with HSG "C"	79
	Approx. 50% Myakka (29) with HSG "B/D"	84
	Approx. 25% St. Johns (46) with HSG "B/D"	84

(Open space, fair condition (50%<grass cover<75%)

Balanced CN: 82.75

	SF	ACRE		%	
Existing structure area	0	0		0	
Existing conc. area	0	0		0	
Total impervious area	0	0	0	98	
Existing pond area	0	0		0	100
Total pervious area	209088	4.80		100	82.75
Total project area (pre-development)	209088	4.80		100	

Pervious area balanced CN   82.75

25-year, 24-hour rainfall depth   7.5-inch

Time of concentration "Tc":

Hydraulic length = 720 feet @ average slope of 0.0043 ft./ft.

Tc = 55.5 minutes
(See the following page for the SCS Tc Computations)

The above-stated CN values were obtained from Table 2.2 Runoff Curve Numbers, Urban Hydrology for Small Watersheds, Technical Release 55, June 1986. The Hydraulic Length and Average Slope were obtained from Figure 7.5 Pre-development Drainage Plan.

Average slope = (145.1 ft. – 142.0 ft.) / 720 ft. = 0.0043 ft./ft.

From the following Figure 15.4 velocity versus slope for shallow concentrated flow (Ref. Chapter 15, National Engineering Handbook), the velocity is approximately 0.216 ft. per Second.

Now, we can compute the time of concentration based on the following formula:

Tc = L / 3600 V Ref. Urban Hydrology for Small Watersheds, Chapter 3: Time of Concentration and Travel Time, Technical Release 55, June 1986.

Where-

Tc = Travel time (hr.),
L = Flow length (ft.),
V = Average velocity (ft./sec.), and
3600 = Conversion factor from seconds to hours.

For the pre-development drainage basin,

Tc = (720 ft.) / (3600) x 0.216 ft./sec. = 0.925 hrs. = 55.5 min.

The following pages are excerpts from the Technical Release 55, Urban Hydrology for the Small Watersheds. The first page (Appendix A) describes four hydrologic soil groups (HSG), A, B, C, and D. There are additional pages (Exhibit A-1) in the same reference which provide a list of soil types and their respective HSG within the United States. Exhibit A-1 is over forty pages long, and that is why we did not include it here.

The following four pages, Table 2-21 to Table 2-2d that follow Appendix A, provide CN values for each soil type and associated HSG in the United States.

Figure 15–4   Velocity versus slope for shallow concentrated flow

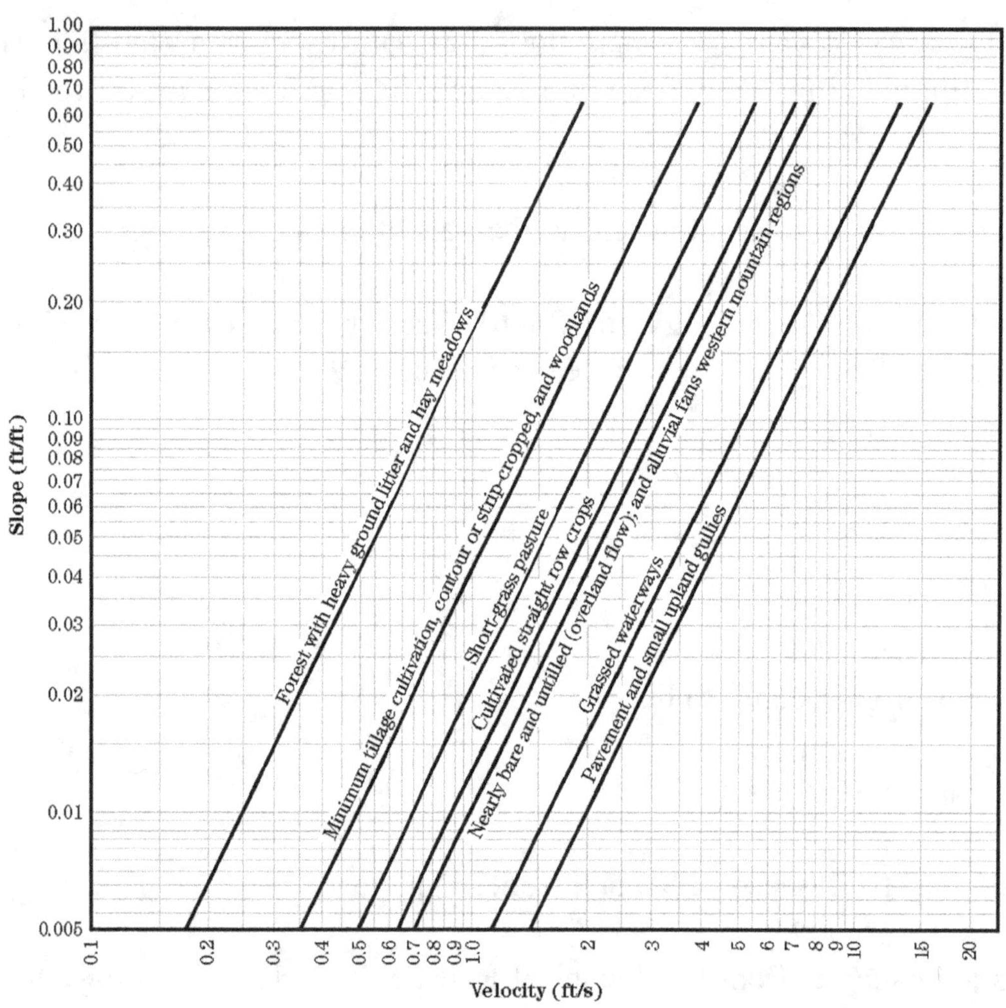

Table 15–3   Equations and assumptions developed from figure 15–4

Flow type	Depth (ft)	Manning's n	Velocity equation (ft/s)
Pavement and small upland gullies	0.2	0.025	$V = 20.328(s)^{0.5}$
Grassed waterways	0.4	0.050	$V = 16.135(s)^{0.5}$
Nearly bare and untilled (overland flow); and alluvial fans in western mountain regions	0.2	0.051	$V = 9.965(s)^{0.5}$
Cultivated straight row crops	0.2	0.058	$V = 8.762(s)^{0.5}$
Short-grass pasture	0.2	0.073	$V = 6.962(s)^{0.5}$
Minimum tillage cultivation, contour or strip-cropped, and woodlands	0.2	0.101	$V = 5.032(s)^{0.5}$
Forest with heavy ground litter and hay meadows	0.2	0.202	$V = 2.516(s)^{0.5}$

Appendix A   Hydrologic Soil Groups

Soils are classified into hydrologic soil groups (HSG's) to indicate the minimum rate of infiltration obtained for bare soil after prolonged wetting. The HSG's, which are A, B, C, and D, are one element used in determining runoff curve numbers (see chapter 2). For the convenience of TR-55 users, exhibit A-1 lists the HSG classification of United States soils.

The infiltration rate is the rate at which water enters the soil at the soil surface. It is controlled by surface conditions. HSG also indicates the transmission rate—the rate at which the water moves within the soil. This rate is controlled by the soil profile. Approximate numerical ranges for transmission rates shown in the HSG definitions were first published by Musgrave (USDA 1955). The four groups are defined by SCS soil scientists as follows:

Group A soils have low runoff potential and high infiltration rates even when thoroughly wetted. They consist chiefly of deep, well to excessively drained sand or gravel and have a high rate of water transmission (greater than 0.30 in/hr).

Group B soils have moderate infiltration rates when thoroughly wetted and consist chiefly of moderately deep to deep, moderately well to well drained soils with moderately fine to moderately coarse textures. These soils have a moderate rate of water transmission (0.15-0.30 in/hr).

Group C soils have low infiltration rates when thoroughly wetted and consist chiefly of soils with a layer that impedes downward movement of water and soils with moderately fine to fine texture. These soils have a low rate of water transmission (0.05-0.15 in/hr).

Group D soils have high runoff potential. They have very low infiltration rates when thoroughly wetted and consist chiefly of clay soils with a high swelling potential, soils with a permanent high water table, soils with a claypan or clay layer at or near the surface, and shallow soils over nearly impervious material. These soils have a very low rate of water transmission (0-0.05 in/hr).

In exhibit A-1, some of the listed soils have an added modifier; for example, "Abrazo, gravelly." This refers to a gravelly phase of the Abrazo series that is found in SCS soil map legends.

Disturbed soil profiles

As a result of urbanization, the soil profile may be considerably altered and the listed group classification may no longer apply. In these circumstances, use the following to determine HSG according to the texture of the new surface soil, provided that significant compaction has not occurred (Brakensiek and Rawls 1983).

HSG	Soil textures
A	Sand, loamy sand, or sandy loam
B	Silt loam or loam
C	Sandy clay loam
D	Clay loam, silty clay loam, sandy clay, silty clay, or clay

Drainage and group D soils

Some soils in the list are in group D because of a high water table that creates a drainage problem. Once these soils are effectively drained, they are placed in a different group. For example, Ackerman soil is classified as A/D. This indicates that the drained Ackerman soil is in group A and the undrained soil is in group D.

Chapter 2	Estimating Runoff	Technical Release 55 Urban Hydrology for Small Watersheds

Table 2-2a Runoff curve numbers for urban areas [1]

Cover description		Curve numbers for hydrologic soil group			
Cover type and hydrologic condition	Average percent impervious area [2]	A	B	C	D
Fully developed urban areas (vegetation established)					
Open space (lawns, parks, golf courses, cemeteries, etc.) [3]:					
Poor condition (grass cover < 50%)		68	79	86	89
Fair condition (grass cover 50% to 75%)		49	69	79	84
Good condition (grass cover > 75%)		39	61	74	80
Impervious areas:					
Paved parking lots, roofs, driveways, etc.					
(excluding right-of-way)		98	98	98	98
Streets and roads:					
Paved; curbs and storm sewers (excluding					
right-of-way)		98	98	98	98
Paved; open ditches (including right-of-way)		83	89	92	93
Gravel (including right-of-way)		76	85	89	91
Dirt (including right-of-way)		72	82	87	89
Western desert urban areas:					
Natural desert landscaping (pervious areas only) [4]		63	77	85	88
Artificial desert landscaping (impervious weed barrier,					
desert shrub with 1- to 2-inch sand or gravel mulch					
and basin borders)		96	96	96	96
Urban districts:					
Commercial and business	85	89	92	94	95
Industrial	72	81	88	91	93
Residential districts by average lot size:					
1/8 acre or less (town houses)	65	77	85	90	92
1/4 acre	38	61	75	83	87
1/3 acre	30	57	72	81	86
1/2 acre	25	54	70	80	85
1 acre	20	51	68	79	84
2 acres	12	46	65	77	82
Developing urban areas					
Newly graded areas					
(pervious areas only, no vegetation) [5]		77	86	91	94
Idle lands (CN's are determined using cover types					
similar to those in table 2-2c).					

[1] Average runoff condition, and $I_a = 0.2S$.

[2] The average percent impervious area shown was used to develop the composite CN's. Other assumptions are as follows: impervious areas are directly connected to the drainage system, impervious areas have a CN of 98, and pervious areas are considered equivalent to open space in good hydrologic condition. CN's for other combinations of conditions may be computed using figure 2-3 or 2-4.

[3] CN's shown are equivalent to those of pasture. Composite CN's may be computed for other combinations of open space cover type.

[4] Composite CN's for natural desert landscaping should be computed using figures 2-3 or 2-4 based on the impervious area percentage (CN = 98) and the pervious area CN. The pervious area CN's are assumed equivalent to desert shrub in poor hydrologic condition.

[5] Composite CN's to use for the design of temporary measures during grading and construction should be computed using figure 2-3 or 2-4 based on the degree of development (impervious area percentage) and the CN's for the newly graded pervious areas.

Table 2-2b   Runoff curve numbers for cultivated agricultural lands [1]

Cover description			Curve numbers for hydrologic soil group			
Cover type	Treatment [2]	Hydrologic condition [3]	A	B	C	D
Fallow	Bare soil	—	77	86	91	94
	Crop residue cover (CR)	Poor	76	85	90	93
		Good	74	83	88	90
Row crops	Straight row (SR)	Poor	72	81	88	91
		Good	67	78	85	89
	SR + CR	Poor	71	80	87	90
		Good	64	75	82	85
	Contoured (C)	Poor	70	79	84	88
		Good	65	75	82	86
	C + CR	Poor	69	78	83	87
		Good	64	74	81	85
	Contoured & terraced (C&T)	Poor	66	74	80	82
		Good	62	71	78	81
	C&T + CR	Poor	65	73	79	81
		Good	61	70	77	80
Small grain	SR	Poor	65	76	84	88
		Good	63	75	83	87
	SR + CR	Poor	64	75	83	86
		Good	60	72	80	84
	C	Poor	63	74	82	85
		Good	61	73	81	84
	C + CR	Poor	62	73	81	84
		Good	60	72	80	83
	C&T	Poor	61	72	79	82
		Good	59	70	78	81
	C&T + CR	Poor	60	71	78	81
		Good	58	69	77	80
Close-seeded or broadcast legumes or rotation meadow	SR	Poor	66	77	85	89
		Good	58	72	81	85
	C	Poor	64	75	83	85
		Good	55	69	78	83
	C&T	Poor	63	73	80	83
		Good	51	67	76	80

[1] Average runoff condition, and $I_a=0.2S$

[2] Crop residue cover applies only if residue is on at least 5% of the surface throughout the year.

[3] Hydraulic condition is based on combination factors that affect infiltration and runoff, including (a) density and canopy of vegetative areas, (b) amount of year-round cover, (c) amount of grass or close-seeded legumes, (d) percent of residue cover on the land surface (good $\geq$ 20%), and (e) degree of surface roughness.

Poor: Factors impair infiltration and tend to increase runoff.

Good: Factors encourage average and better than average infiltration and tend to decrease runoff.

Table 2-2c   Runoff curve numbers for other agricultural lands [1]

Cover description		Curve numbers for hydrologic soil group			
Cover type	Hydrologic condition	A	B	C	D
Pasture, grassland, or range—continuous forage for grazing. [2]	Poor	68	79	86	89
	Fair	49	69	79	84
	Good	39	61	74	80
Meadow—continuous grass, protected from grazing and generally mowed for hay.	—	30	58	71	78
Brush—brush-weed-grass mixture with brush the major element. [3]	Poor	48	67	77	83
	Fair	35	56	70	77
	Good	30 [4]	48	65	73
Woods—grass combination (orchard or tree farm). [5]	Poor	57	73	82	86
	Fair	43	65	76	82
	Good	32	58	72	79
Woods. [6]	Poor	45	66	77	83
	Fair	36	60	73	79
	Good	30 [4]	55	70	77
Farmsteads—buildings, lanes, driveways, and surrounding lots.	—	59	74	82	86

[1] Average runoff condition, and $I_a = 0.2S$.

[2] *Poor:* <50%) ground cover or heavily grazed with no mulch.
Fair: 50 to 75% ground cover and not heavily grazed.
Good: > 75% ground cover and lightly or only occasionally grazed.

[3] *Poor:* <50% ground cover.
Fair: 50 to 75% ground cover.
Good: >75% ground cover.

[4] Actual curve number is less than 30; use CN = 30 for runoff computations.

[5] CN's shown were computed for areas with 50% woods and 50% grass (pasture) cover. Other combinations of conditions may be computed from the CN's for woods and pasture.

[6] *Poor:* Forest litter, small trees, and brush are destroyed by heavy grazing or regular burning.
Fair: Woods are grazed but not burned, and some forest litter covers the soil.
Good: Woods are protected from grazing, and litter and brush adequately cover the soil.

Table 2-2d Runoff curve numbers for arid and semiarid rangelands [1]

Cover description		Curve numbers for hydrologic soil group			
Cover type	Hydrologic condition [2]	A [3]	B	C	D
Herbaceous—mixture of grass, weeds, and	Poor		80	87	93
low-growing brush, with brush the	Fair		71	81	89
minor element.	Good		62	74	85
Oak-aspen—mountain brush mixture of oak brush,	Poor		66	74	79
aspen, mountain mahogany, bitter brush, maple,	Fair		48	57	63
and other brush.	Good		30	41	48
Pinyon-juniper—pinyon, juniper, or both;	Poor		75	85	89
grass understory.	Fair		58	73	80
	Good		41	61	71
Sagebrush with grass understory.	Poor		67	80	85
	Fair		51	63	70
	Good		35	47	55
Desert shrub—major plants include saltbush,	Poor	63	77	85	88
greasewood, creosotebush, blackbrush, bursage,	Fair	55	72	81	86
palo verde, mesquite, and cactus.	Good	49	68	79	84

[1] Average runoff condition, and $I_a = 0.2S$. For range in humid regions, use table 2-2c.
[2] Poor: <30% ground cover (litter, grass, and brush overstory).
   Fair: 30 to 70% ground cover.
   Good: > 70% ground cover.
[3] Curve numbers for group A have been developed only for desert shrub.

7.10. Post-development Site Conditions

We proposed designing and constructing approximately 5,600 feet of paved roadway to serve sixteen proposed residential lots of 10,000 SF (+/-). We assumed that each lot would have about 2,200 SF of residential homes, including related impervious surfaces such as driveways and sidewalks.

Curbs and gutters with storm pipes and inlets will be constructed to collect and convey storm-water runoff to the proposed dry detention ponds located on both sides of the proposed site entrance. As in pre-development site conditions, the final discharge point of the site will remain intact.

The post-development condition's project area is subdivided into four smaller sub-drainage areas to facilitate stormwater design and computations.

(See Figure 7-6 for the post-development drainage plan on the next page)

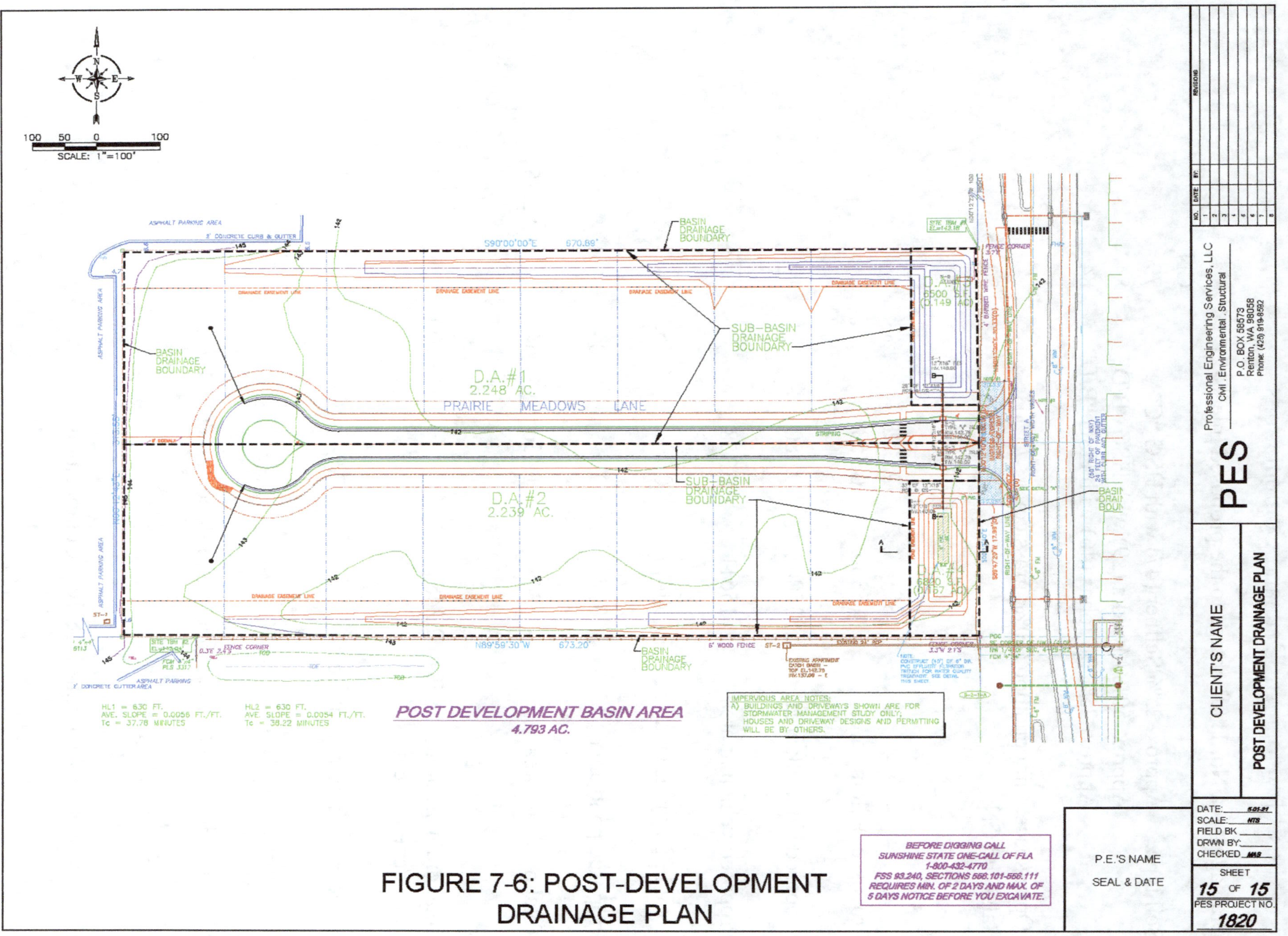

FIGURE 7-6: POST-DEVELOPMENT DRAINAGE PLAN

7.11. Post-development Data and Computations

Drainage Area No. 1: (discharge to the proposed Inlet S-2)

		CN
Soil type:	Approx. 25% Seffner (47) with HSG "C"	79
	Approx. 50% Myakka (29) with HSG "B/D"	84
	Approx. 25% St. Johns (46) with HSG "B/D"	84

(Open space, fair condition (50%<grass cover<75%)
Balanced CN: 82.75

	SF	ACRE	%	
Proposed building area	15360	0.353		
Proposed sidewalk area	3506	0.080		
Proposed vehicular area	15611	0.358		
Total impervious area	34477	0.791	35.2	98
Proposed pond area	0	0	0	98
Total pervious area	63436	1.456	64.8	82.75

Total drainage area - 1 97913 2.248   100
(post-development)

25-year, 24-hour rainfall depth   7.5-inch

Time of concentration "Tc-1":

Hydraulic length = 630 feet @ average slope of 0.0056 ft./ft.

Tc-1 = 37.78 minutes

(Follow Section 7.9 for the SCS Tc-1 Computation)

Drainage Area No. 2: (discharge to the proposed Inlet S-3)

		CN
Soil type:	Approx. 25% Seffner (47) with HSG "C"	79
	Approx. 50% Myakka (29) with HSG "B/D"	84
	Approx. 25% St. Johns (46) with HSG "B/D"	84

(Open space, fair condition (50%<grass cover<75%)

Balanced CN:				82.75

	SF	ACRE	%	
Proposed building area	15360	0.353		
Proposed sidewalk area	3507	0.081		
Proposed vehicular area	15611	0.358		
Total impervious area	34478	0.792	35.3	98
Proposed pond area	0	0	0	98
Total pervious area	63071	1.448	64.7	82.75
Total drainage area - 2 (post-development)	97549	2.239	100	

25-year, 24-hour rainfall depth   7.5-inch

Time of concentration "Tc-2":

Hydraulic length = 630 feet @ average slope of 0.0054 ft./ft.

Tc-2 = 38.22 minutes

(Follow Section 7.9 for the SCS Tc-2 Computation)

Drainage Area No. 3: (discharge to the proposed Pond 1)

		CN
Soil type:	Approx. 25% Seffner (47) with HSG "C"	79
	Approx. 50% Myakka (29) with HSG "B/D"	84
	Approx. 25% St. Johns (46) with HSG "B/D"	84

(Open space, fair condition (50%<grass cover<75%)
Balanced CN: 82.75

	SF	ACRE	%	
Proposed building area	0	0		
Proposed sidewalk area	0	0		
Proposed vehicular area	0	0		
Total impervious area	0	0	0	98
Proposed pond area	4999	0.115	76.9	98
Total pervious area	1501	0.034	23.1	82.75
Total drainage area - 3 (post-development)	6500	0.149	100	

Pervious area balanced CN 94

25-year, 24-hour rainfall depth 7.5-inch

Time of concentration "Tc-3" = 10 minutes

We are assuming a ten-minute time of concentration since the drainage area is small. The actual value of the Tc for this small drainage area is perhaps lower than the ten minutes.

Drainage Area No. 4: (discharge to the proposed Pond 2)

		CN
Soil type:	Approx. 25% Seffner (47) with HSG "C"	79
	Approx. 50% Myakka (29) with HSG "B/D"	84
	Approx. 25% St. Johns (46) with HSG "B/D"	84

(Open space, fair condition (50%<grass cover<75%)

Balanced CN: 82.75

	SF	ACRE	%	CN
Proposed building area	0	0		
Proposed sidewalk area	0	0		
Proposed vehicular area	0	0		
Total impervious area	0	0	0	98
Proposed pond area	5209	0.120	76.4	98
Total pervious area	1611	0.037	23.6	82.75

Total drainage area - 3   6820 0.157 100
(post-development)

Pervious area balanced CN   94

25-year, 24-hour rainfall depth   7.5-inch

Time of concentration "Tc-3" = 10 minutes

We are assuming a ten-minute time of concentration since the drainage area is small. The actual value of the Tc for this small drainage area is perhaps lower than the ten minutes.

Total post-development project area   208782 SF (4.793 Acres)

Total post-development impervious area   68955 SF (1.583 Acres)

7.12. Data Analysis and Comparison

A. Water Quality Requirements

Project area = 208782 SF (4.793 Acre)

Volume required (dry pond) = 208782 SF x 0.5 inch / 12 (inch/foot) = 8699 CF

B. Seasonal High-Water Elevation (SHWE) Evaluation:

The following data was obtained from the geotechnical report the geotechnical engineering firm provided:

SB – 1	feet
Approximate ground elevation	142.00
Seasonal high-water depth	7.00
Seasonal high-water elevation (SHWE)	135.00

SB – 2	feet
Approximate ground elevation	142.00
Seasonal high-water depth	7.00
Seasonal high-water elevation (SHWE)	135.00

We assumed a conservative value for SHWE   137.00

C. Ponds Stage Storage Data:

Stage (FT)	Elev. (FT)	Pond 1 Area (SF)	Pond 1 Area (Acre)	Pond 2 Area (SF)	Pond 2 Area (Acre)	Total Area (SF)	Volume (CF)	Acc. Volume (CF)	Remark
140	0.00	2728	0.063	2909	0.067	5637.000	0	0	BOP
140.25	0.25	2890.21	0.066	3073.28	0.071	5963.490	1450	1450	
140.50	0.50	3052.42	0.070	3237.56	0.074	6289.980	1532	2982	
140.75	0.75	3214.63	0.074	3401.84	0.078	6616.470	1613	4595	
141.00	1.00	3376.84	0.078	3566.12	0.082	6942.960	1695	6290	
141.25	1.25	3539.05	0.081	3730.4	0.086	7269.450	1777	8067	
141.50	1.50	3701.26	0.085	3894.68	0.089	7595.940	1858	9925	Weir
141.75	1.75	3863.47	0.089	4058.96	0.093	7922.430	1940	11865	Crest
142.00	2.00	4025.68	0.092	4223.24	0.097	8248.920	2021	13886	
142.25	2.25	4187.89	0.096	4387.52	0.101	8575.410	2103	15989	
142.50	2.50	4350.1	0.100	4551.8	0.104	8901.900	2185	18174	
142.75	2.75	4512.31	0.104	4716.08	0.108	9228.390	2266	20440	
143.00	3.00	4674.52	0.107	4880.36	0.112	9554.880	2348	22788	
143.25	3.25	4836.73	0.111	5044.64	0.116	9881.370	2430	25217	
143.50	3.50	4998.94	0.115	5208.92	0.120	10207.860	2511	27729	TOP

Weir width = 2.0 ft.

Volume provided = 9925 CF (+14%) O.K.

Above stated volume will be provided between BOP @ 140-foot and the Weir Crest Elevation @ 141.50-foot.

 D. Recovery Time

According to the Natural Resources Conservation Service, soil type: Myakka Fine Sand or Seffner Fine Sand, permeability rate of 6.0 to 20.0 inch/hour.

Assumed average; Kv = 13.0 inch/hour x 2.0 = 26.0 feet/day (conservative)
Kh = 26.0 x 1.5 = 39.0 feet/day
Approximate pond length = 216 feet
Approximate pond width = 53 feet

As shown on the next page, mounding analysis, pond(S) recover @63.3 hours less than the allowable 72.0 hours. See the following pages for input and output data.

 E. Peak Discharge

25-year, 24-hour post-development peak discharge of 7.29 CFS is less than the pre-development peak discharge of 7.59 CFS.

 F. Free Board

About eight-inch freeboard is provided above the design high water elevation of 142.85 feet.

The following pages represent a summary of XPSWMM input and output data for the pre-and post-development, ponds stage-storage, outfall structure, mounding analysis and associated hydrographs. The actual input and output data are voluminous and not presented here.

Pre-Development Runoff Study, 25-Year – 24-Hour
#######################################
Rainfall input summary from runoff
#######################################
Total rainfall for gage # 1 is 7.5000 inches
##
Table R3. SUBCATCHMENT DATA
##
1 PRE-DEV.#1 1 SCS Method SCS curvilinear

Total number of sub catchments ...   1
Total tributary area (acres) ...   4.79
Impervious area (acres) 0.00
Pervious area (acres) 4.79
Total width (feet) 1.00
Percent imperviousness 0.00

Pre-Development Runoff Study, 25-Year – 24-Hour
PRE-DEV.
Date Time   Flow
Mo/Da/Year Hr:Min   Cfs
---------- ------ ------
1 1 2007 0 6   0.0000
Maximum value   **7.5910**
Minimum value   0.0000   -
Total loads   9.14E+04 cubic-ft
===> Runoff simulation ended normally.
===> Hydraulic model simulation ended normally.

Post-Development Runoff Study, 25-Year – 24-Hour
######################################
Rainfall input summary from runoff
######################################
Total rainfall for gage # 1 is 7.5000 inches
##
Table R3. SUBCATCHMENT DATA
##

1 S-3#1   1 SCS Method SCS curvilinear SCS Type II Mod.1.1
2 Pond 2#1 1 SCS Method SCS curvilinear SCS Type II Mod.1.1
3 S-2#1   1 SCS Method SCS curvilinear SCS Type II Mod.1.1
4 Pond 1#1 1 SCS Method SCS curvilinear SCS Type II Mod.1.1
Total number of sub catchments...   4
Total tributary area (acres)....   4.79
Impervious area (acres).........   1.58
Pervious area (acres)...........   3.21
Total width (feet)..............   4.00
Percent imperviousness..........   33.00
****
* Summary of quantity results (flow in cfs) *
****

Post-Development Runoff Study, 25-Year – 24-Hour

	S-3	Pond 2	S-2	Pond 1
Date Time	Flow	Flow	Flow	Flow
Mo/Da/Year Hr:Min	Cfs	Cfs	Cfs	Cfs
1 1 2007 0 6	0.0000	0.0000	0.0000	0.0000
Maximum value	**4.8399**	**0.6004**	**4.9054**	**0.5698**
Minimum value	0.0000	0.0000	0.0000	0.0000
Total loads	4.75E+04	3.52E+03	4.79E+04	3.34E+03
	Cubic-ft	Cubic-ft	Cubic-ft	Cubic-ft

===> Runoff simulation ended normally.

Post Development Hydraulic Study 25-Year – 24-Hour

==

| Table E1 - Conduit Data |

==

Trapezoid

Inp Num	Conduit Name	Length (ft)	Conduit Class	Area (ft^2)	Manning Coef.	Max (ft)	Width (ft)	Depth	Side Slopes
1	P 3-Pond 2	32.0000	Circular	1.7671	0.0140	1.5000	1.5000		
2	P 2-3	42.0000	Circular	1.7671	0.0140	1.5000	1.5000		
3	P 4-Pond 1	28.0000	Circular	1.7671	0.0140	1.5000	1.5000		

Total length of all conduits .... 102.0000 feet

=====================

| Conduit Volume |

=====================

Full pipe or full open conduit volume
Input full depth volume............. 1.8378E+02 cubic feet

==

| Table E3a - Junction Data |

==

Inp Num	Junction Name	Ground Elevation	Crown Elevation	Invert Elevation	Qinst cfs	Initial Depth-ft	Interface Flow (%)
1	Pond 2	143.0000	143.0000	140.0000	0.0000	0.0000	100.0000
2	S-3	142.7800	142.7800	140.0000	0.0000	0.0000	100.0000
3	Outfall	143.0000	143.0000	141.0000	0.0000	0.0000	100.0000
4	S-2	142.7800	142.7800	140.0000	0.0000	0.0000	100.0000
5	Pond 1	143.0000	143.0000	140.0000	0.0000	0.0000	100.0000

==

| Table E3b - Junction Data |

==

Inp Num	Junction Name	X Coord.	Y Coord.	Type of Manhole	Type of Inlet	Maximum Capacity
1	Pond 2	90.5685	405.9924	Flooded ponding	Normal inlet	
2	S-3	90.6572	421.3229	Flooded ponding	Normal inlet	
3	Outfall	90.5196	393.4740	Flooded ponding	Normal inlet	
4	S-2	90.5179	433.9883	Flooded ponding	Normal inlet	
5	Pond 1	90.6402	445.9597	Flooded ponding`	Normal inlet	

==

| Table E4 - Conduit Connectivity |

==

Input Number	Conduit Name	Upstream Node	Downstream Node	Upstream Elevation	Downstream Elevation
1	P 3-Pond 2	S-3	Pond 2	140.0000	140.0000
2	P 2-3	S-3	S-2	140.0000	140.0000
3	P 4-Pond 1	S-2	Pond 1	140.0000	140.0000

====================================

| Variable storage data for node |Pond 2

====================================

Data Point	Elevation ft	Depth ft	Area ft^2	Volume ft^3	Area acres	Volume ac-ft
1	140.0000	0.0000	2918.5200	0.0000	0.0670	0.0000
2	140.5000	0.5000	3223.4400	1534.8589	0.0740	0.0352
3	141.0000	1.0000	3571.9200	3232.9538	0.0820	0.0742
4	141.5000	1.5000	3876.8400	5094.6235	0.0890	0.1170
5	142.0000	2.0000	4225.3200	7119.5386	0.0970	0.1634
6	142.5000	2.5000	4530.2400	9307.9861	0.1040	0.2137
7	143.0000	3.0000	4878.7200	11659.6881	0.1120	0.2677

====================================

| Variable storage data for node |Pond 1

=======================================

Data Point	Elevation ft	Depth ft	Area ft^2	Volume ft^3	Area acres	Volume ac-ft
1	140.0000	0.0000	2744.2800	0.0000	0.0630	0.0000
2	140.5000	0.5000	3049.2000	1447.7009	0.0700	0.0332
3	141.0000	1.0000	3397.6800	3058.6354	0.0780	0.0702
4	141.5000	1.5000	3702.6000	4833.1595	0.0850	0.1110
5	142.0000	2.0000	4007.5200	6760.1869	0.0920	0.1552
6	142.5000	2.5000	4356.0000	8850.4616	0.1000	0.2032
7	143.0000	3.0000	4660.9200	11104.2619	0.1070	0.2549

=======================================

| Weir Data |

=======================================

Weir Name	From Junction	To Junction	Crest Type	Weir Height(ft)	Weir Top(ft)	Discharge Length(ft)	Weir Coefficient	Power
C.S.1	Pond 2	Outfall	1	1.50	3.00	2.00	3.1300	1.5000

=======================================

| Table E9 - JUNCTION SUMMARY STATISTICS |
| The maximum area is only the area of the node, it |
| does not include the area of the surrounding conduits |

=======================================

Junction Name	Ground Elevation feet	PipeCrown Elevation feet	Junction of Elevation feet	Surcharge Occurrence Hr.	Min.	Freeboard at Max Elevation	Junction of Node feet	Area ft^2
Pond 2	143.0000	143.0000	142.6069	12	49	0.0000	0.3931	4604.7680
S-3	142.7800	141.5000	142.7774	12	47	0.0000	0.0026	12.5660
Outfall	143.0000	141.0000	141.0000	0	0	0.0000	2.0000	12.5660
S-2	142.7800	141.5000	142.8390	13	49	0.0590	0.0000	67139.671
Pond 1	143.0000	143.0000	142.8463	13	53	0.0000	0.1537	4567.2122

=======================================

| Table E21. Continuity balance at the end of the simulation |
| Junction Inflow, Outflow or Street Flooding |
| Error = Inflow + Initial Volume - Outflow - Final Volume |

```
*==================================================================*

     Inflow         Inflow          Average
     Junction       Volume,ft^3     Inflow, cfs
   ---------------   ------------    -------------

     Pond 2          3522.6115       0.0408
     S-3             47419.6036      0.5488
     S-2             47843.9366      0.5537
     Pond 1          3343.1154       0.0387
     Outflow         Outflow         Average
     Junction        Volume,ft^3     Outflow, cfs
   ---------------   ------------    -------------

     Outfall         94299.7318      1.0914
*==================================================================*
| Initial system volume =              0.0010 Cu Ft |
| Total system inflow volume =         102135.4807 Cu Ft |
| Inflow + initial volume =               102135.4817 Cu Ft |
*==================================================================*
| Total system outflow =               94299.7318 Cu Ft |
| Volume left in system =              11529.8249 Cu Ft |
| Evaporation =                        0.0000 Cu Ft |
| Outflow + final volume =             105829.5568 Cu Ft |
*==================================================================*
```

===> Hydraulic model simulation ended normally.

Node - Pond 1
[Max Stage = 142.846]

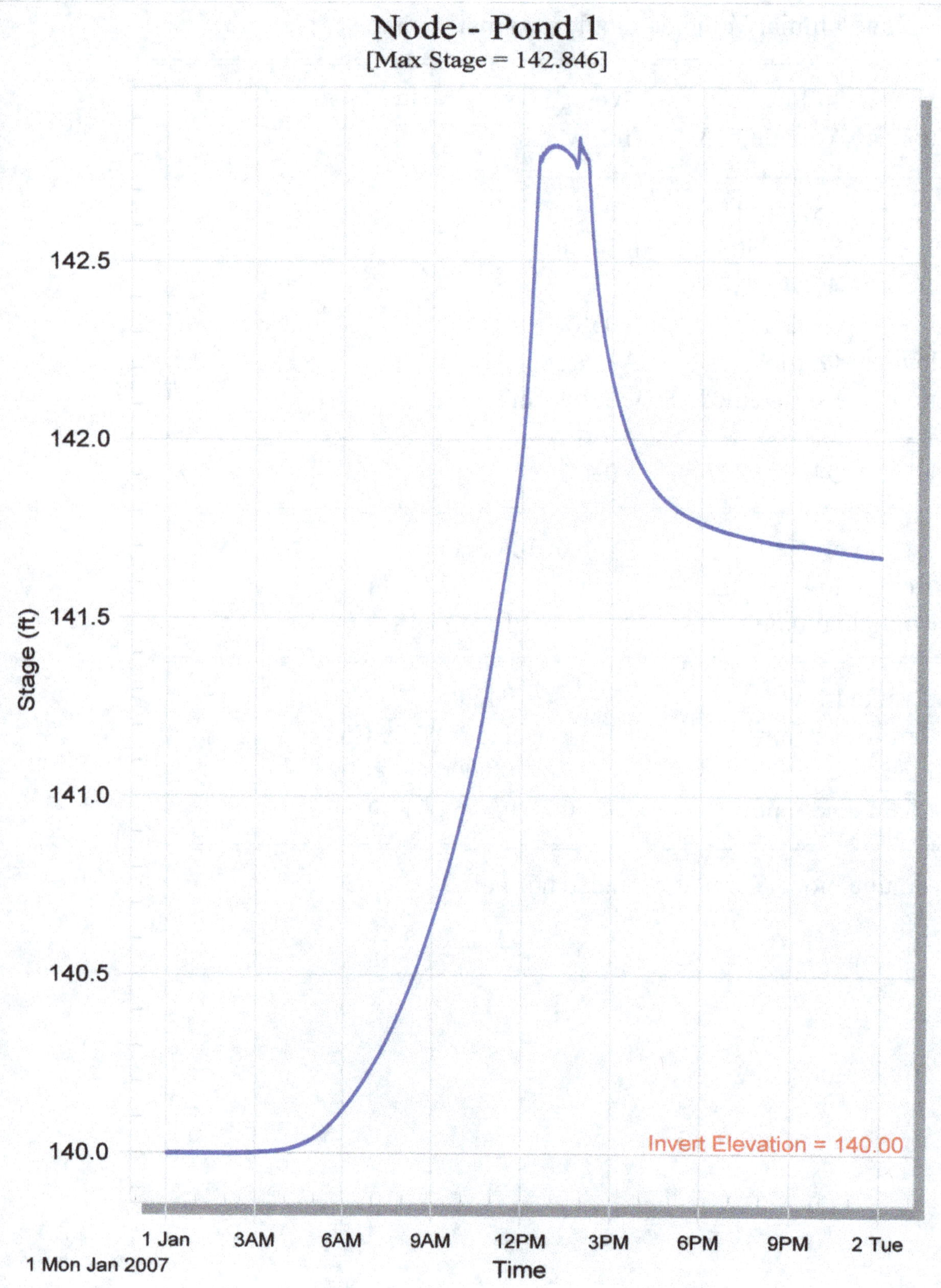

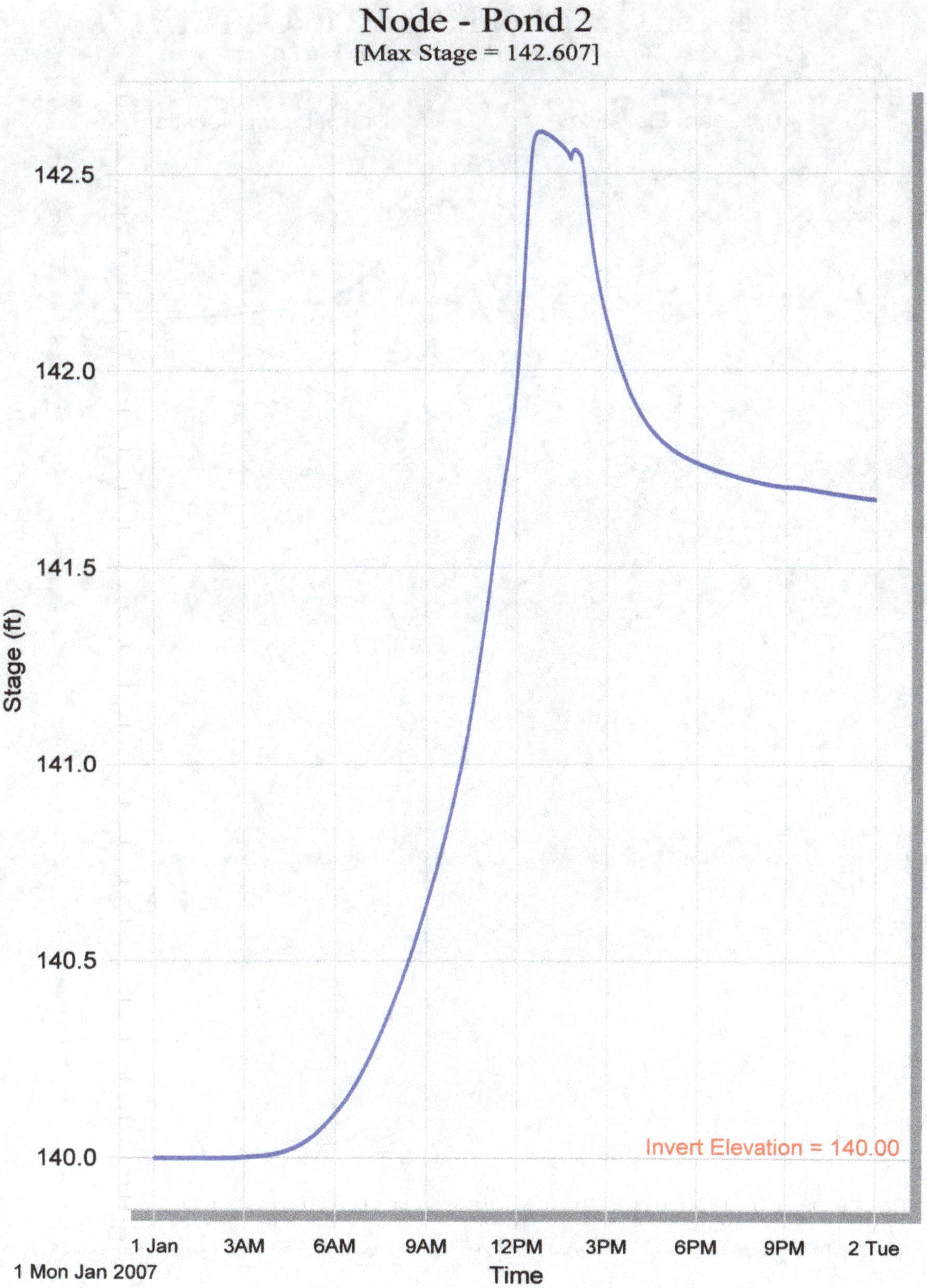
Node - Pond 2
[Max Stage = 142.607]
142.5
142.0
141.5
141.0
140.5
140.0
Stage (ft)
Invert Elevation = 140.00
1 Jan
3AM
6AM
9AM
12PM
3PM
6PM
9PM
2 Tue
1 Mon Jan 2007
Time

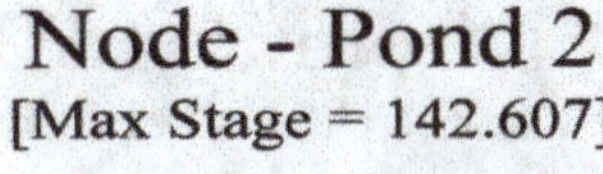

Diversion C.S.1 from Pond 2 to Outfall
[Max Flow = 7.2904][Max Velocity = 0.00]

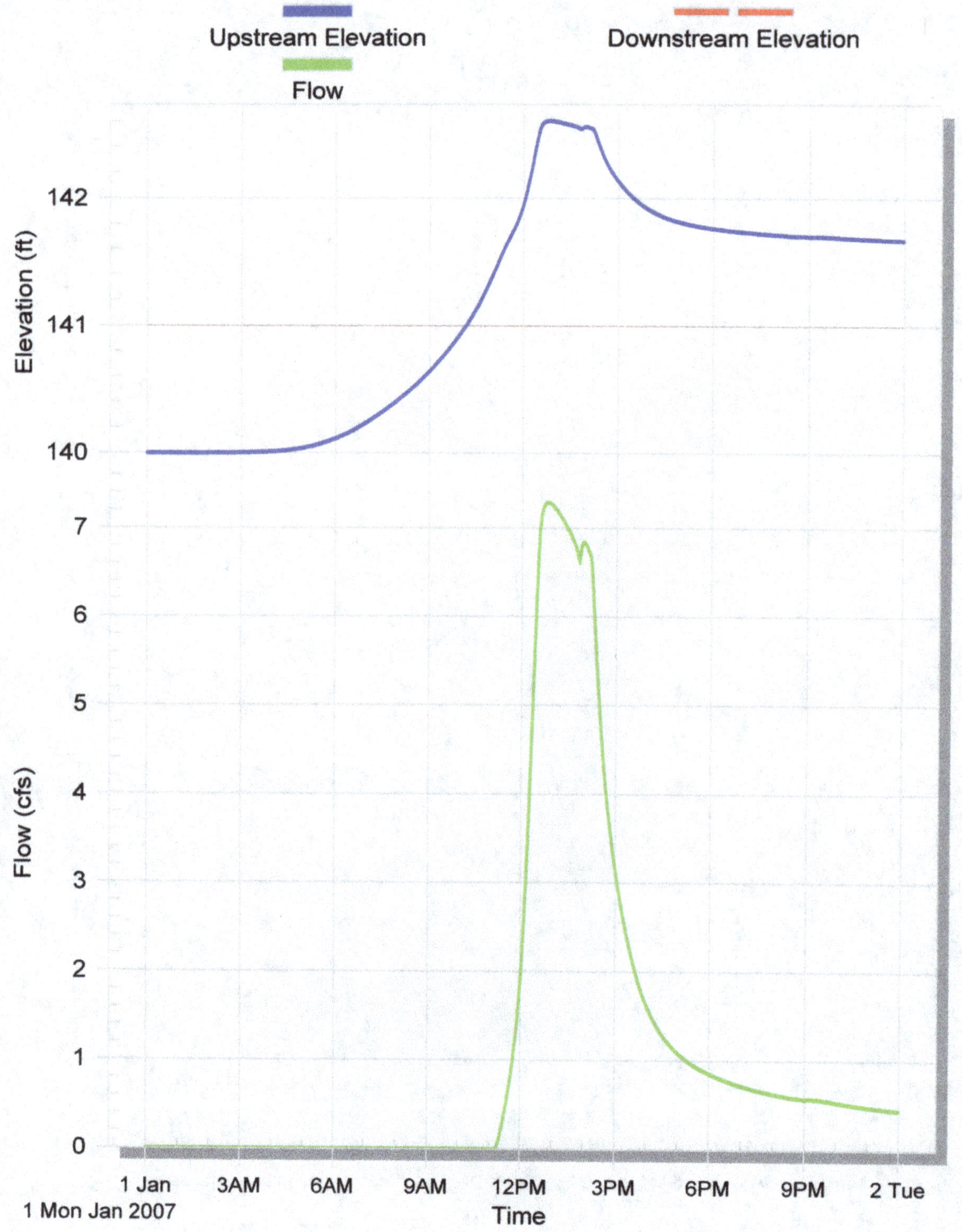

Mounding Analysis - Simplified Analytical Method

Pond Length = L	216	feet
Pond Width = W	53	feet
Porosity = P	1.00	Porosity of material within the pond
Top of Treatment	141.50	feet
SHW	139.00	feet
Bottom of Pond	140.00	feet
Aquifer elevation	136.00	feet
η	0.20	Specific yield or effective porosity of aquifer - see table below
k	39.00	ft/day - horizontal hydraulic conductivity of aquifer
Recovery Time =	63.63	hours
Recovered Volume =	17,172	cubic feet
R_{max} =	185.4	feet - Maximum Radius of Influence
h_{max} =	2.50	feet - Top of treatment to SHW
h_{min} =	1.00	feet - Bottom of pond to SHW
b =	3.00	feet - initial saturated thickness of aquifer

Taken from "
Retention Po
prepared for
by PSI/Jamm
http://sjrwmd

The fillable p
to vary with c

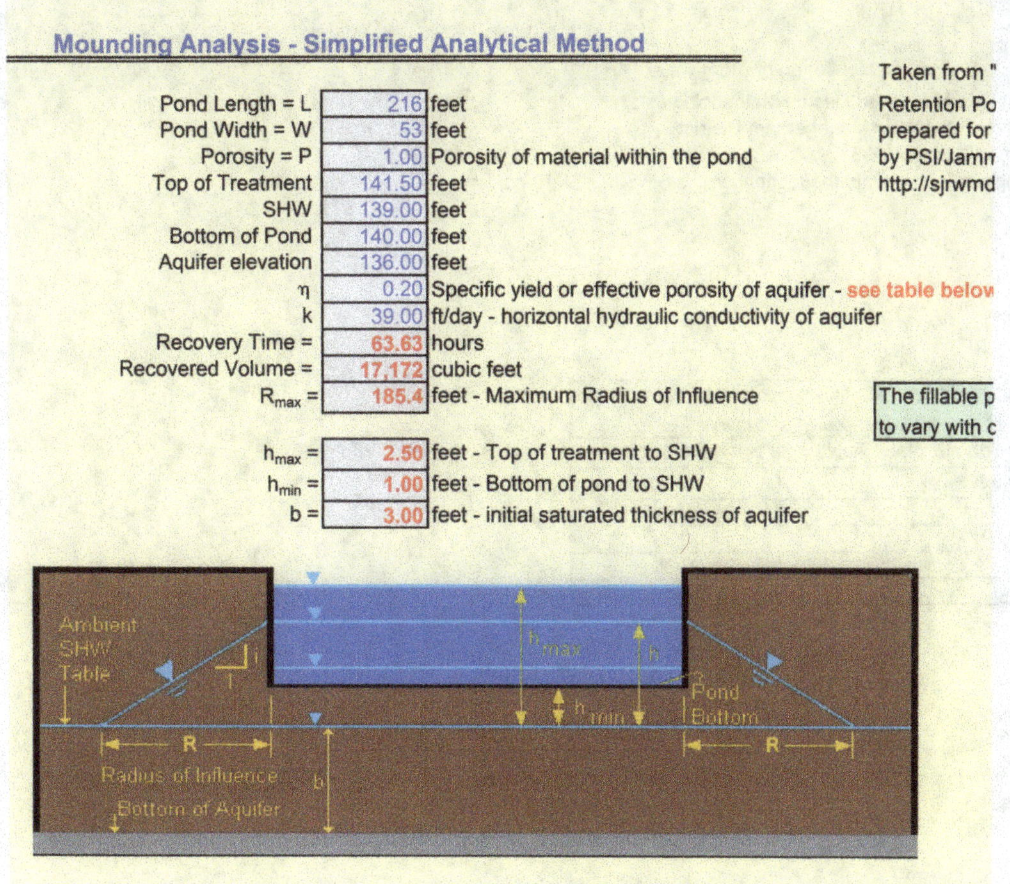

"Full-Scale Hydrologic Monitoring of Stormwater
onds and Recommended Design Methodologies",
St. Johns River Water Management District
nal Division, August 1993.
d.org/programs/outreach/pubs/techpubs/pdfs/SP/SJ93-SP10.pdf

orosity for poorly graded fine sands can be expected
distance above the water table as follows:

Distance above Water Table	Fillable Porosity
0.0 to 0.5 ft	0 %
0.5 to 1.0 ft	10 %
1.0 to 1.5 ft	20 %
1.5 to 2.0 ft	27 %
2.0+ ft	30 %

It is essential to mention here that the computer software used for this project is among many programs on the market today. The author used other programs such as the Western Washington Hydrology Model, and HydroCAD and reached the same conclusion. Many other programs on the market can be used to obtain similar results. Of course, some programs may be more sophisticated than others, but the critical factor is understanding the problem and what we are trying to solve. Be able to formulate the input data and interpret the output data.

7.13. Final Construction Plans Preparation

The following pages explain and include complete construction plan sheets 1 to 15 of 15 for this project. The actual construction plans follow the explanations for all the pages.

1) Sheet 1 of 15 - Cover

Reviewing thousands of projects, we noticed that not every engineer was keen to provide a cover sheet. Often, construction plans are submitted with lots of engineering data and other important information jammed into one or two sheets. It takes much longer for the permitting agencies to review these construction plans. A cover sheet with an index of sheets (even for two or three sheets of construction plans), project title, address, vicinity map, and contact information such as the owner, engineer, architect, and utility companies contact information should be an essential part of any construction plan set.

The index of sheets (like a table of contents of a book) includes each sheet number and corresponding title.

Show the state map of the project on this sheet. Although, this is not necessary for small projects of this size. A location or vicinity map with section, township, and range is more appropriate.

Provide the project title, owners, surveyor, engineer, and the contractor's contact information on this sheet. Utility companies' contact information can also be beneficial to show.

Call before you dig is a vital utility locate phone number that may differ in each state and should be included at least on the cover sheet, but preferably on all construction plan sheets.

Sheet 2 of 15 - General Notes and Specifications

General notes and specifications for the project are provided on this sheet. These notes' primary purpose is to remind the general contractor regarding available information, such as working in the public ROW, the United States Department of Labor Occupational Safety and Health Administration (OSHA) safety requirements, inspection requirements, as-built submittal, etc. A separate document to include project specifications for much larger projects is more appropriate and perhaps necessary. For smaller projects, having all pieces of information as a part of the construction plans might make more sense.

Provide the construction sequence notes on this sheet. Even though the construction sequence may be very similar for the same construction projects, some permitting agencies and their site inspectors may have specific requirements. Also, often contractors have their means and methods of constructing projects, and their way may be a little different from what the professional engineer is proposing. Therefore, it is best to double-check these notes with the AHJ and use them if they have one.

2) Sheet 3 of 15 - Site Survey (Existing Site Conditions)

As mentioned before, the survey data and information, including boundary, topographic, trees, and other environmentally sensitive features, is essential for any successful project. These data set the stage for the rest of the project design, permitting, and construction.

In this project, a professional land surveyor prepared survey sheet was brought into the engineer's title block as sheet 3 of 15 (see Figure 7-1). None of the information on the survey sheet is changed or manipulated in any shape or form. This condition will allow the permitting agencies to look at these data without mingling them with the proposed design to feel the existing site conditions appropriately.

3) Sheet 4 of 15 - Site Plan

The site plan should include both the existing and the proposed information and site data to provide a realistic picture of the site conditions for the AHJ to review. It should also offer sufficient geometric dimensions for structures envelopes, driveways, roadways, curb, gutter, sidewalk, cul-de-sac, landscape islands, and ponds for the surveyor and the contractor to perform the field's horizontal layout.

Zoning information, project site data, building square footage, setbacks, parking information, and flood zone information are also listed under the site data on this sheet.

4) Sheet 5 of 15 - Paving, Grading and Drainage Plan

This sheet may be the more challenging sheet in the entire set to compute correctly for an optimum design. It should have sufficient elevations to define the proposed site topography and contours. It should include structures envelopes, corner elevations, driveways, roadways, curb, gutter, sidewalk, cul-de-sac, landscape islands, and ponds for the surveyor and the contractor to perform the field's vertical layout. For this site, proposed elevations on the east side are controlled by the existing sidewalk elevations. The current parking lot's curb elevations on the west side are the controlling factors. Similarly, we can match the existing elevations on the south and the north sides along the property lines. All other elevations between the property lines are determined by grading the site and how the proposed stormwater sheet flows or is conveyed to the proposed ponds via the storm sewer system.

Two crucial soil profile data indicated on this sheet provide the actual soil properties at the pond locations and the seasonal high-water table (SHWT) at each pond. For a dry pond to function as a dry pond, the SHWT should be located a minimum of two feet below the proposed bottom of the pond elevation. The existing ground elevations at each SB-1 and SB-2 were 142 feet. Geotechnical engineering estimated the SHWT to be about seven feet below the grade. It means the seasonal high-water table was estimated to be at 135 feet. We assumed a more conservative value of 137 feet for the SHWE. We set the bottom of the proposed ponds at an elevation of 140 feet. If the SHWT fluctuates and reaches 138 feet, our pond bottom will still be two feet above this elevation.

The process of designing detention ponds is more like trial and error. The controlling factors are:

- SHWT,
- Pond top and bottom elevations,
- Side slopes (zero for a vault),
- Cross-sectional areas (equals for a vault),
- Square footage of pond at each elevation (or ponds volume),
- Control structure dimensions and elevations,
- Stormwater peak discharge and runoff volume,
- Discharge method.

We set the pond bottom and top elevations to establish the roadway crown elevation of 143.00 feet, Station 1+00 at the entrance (see cross-sections A-A and B-B as shown on sheet 6 of 15, cross sections). Then, we proceed along the centerline of the roadway and establish proposed elevations every fifty feet. The proposed roadway profiles are shown on sheet 8 of 15 (sanitary profile). We raise the roadway elevation 0.2 feet for every fifty feet (0.4% slope) uphill toward the cul-de-sac.

Note: This site is reasonably flat, and the slope values are kept to a minimum value to balance the amount of site excavation (cut and fill). For a steep site, the slope values could be much higher.

Let us review some of the information provided on Sheet 5 of 15.

Erosion Control During Construction Note:

This note provides basic guidelines and instructions to the contractor to control erosion and sediments during construction.

Stormwater System Operation and Management Note:

This note provides guidelines and instructions to the owner or operator. In this case, the HOA will operate and maintain the specific stormwater management system. This practice is necessary for every site to ensure that the constructed facilities function as the design intended. However, it is also a requirement in some states, such as Florida, requiring the EOR to inspect and certify the stormwater management system and its functionality to the water management district, which issued a permit for the project.

Control Structure (Flume) Detail:

This detail at the northeast end of Pond 1 allows the pond to overflow and discharge into the existing ditch along the roadway. The discharge amount is based on the computation provided earlier in this chapter.

Detail "A":

This detail shows the connection between the infiltration system at the bottom of Pond 2 and the existing storm sewer pipe located along the external roadway. This infiltration system aims to allow additional discharge to the ground to compensate for the pond's small bottom area.

We could have increased the pond bottoms to eliminate the infiltration system. However, we had to reduce either the number of lots or at least the size of two adjacent lots, 1 and 15. This decision was based on experience and was an engineering judgment call.

5) Sheet 6 of 15 - Cross Sections

This sheet is a continuation of the previous sheet 5 of 15. Cross-sections referenced on the Paving, Grading, and Drainage plan sheet are shown on this sheet.

Section A-A:

This section provides information on the seasonal high-water elevation, pond's top, bottom, and side slopes. This cross-section also provides significant elevation, such as design high-water elevation (DHWE). The DHWE shows how high the water inside the pond reaches for a particular storm event, in this case, a twenty-five-year storm event.

It is essential to provide information and elevations regarding this cross-section, the existing ditch, and the roadway. We should show how the proposed section transition into the current conditions. In this case, a ditch and then the existing street are located on the proposed pond's east side.

As discussed in Chapter 2, we should emphasize how important it is to obtain accurate survey data regarding the existing site conditions. Otherwise, properly transitioning from the proposed to existing conditions will become erroneous.

Section B-B:

This section provides information and data about pond A and pond B relations. In this case, the connection is a simple pipe (equalizer) connection to equalize water elevations in two ponds.

It also indicated the cover over the proposed pipes, inlet locations, elevations on both sides of the proposed street, the landscape island, pipes sizes, materials, slopes, SHWT, and DHWE.

The above cross-sections also provide notes regarding compaction requirements for the berms.

Street A:

This sanitary sewer profile is shown on this sheet since there wasn't enough space on Sanitary Profile Sheet 8 of 15. It shows the proposed manhole 1 connecting to the existing manhole S-2-1-A. It also includes the designed pipe material, slope, manholes rims, and inverts.

6) Sheet 7 of 16 - Utility and Traffic Plan and Sheet 8 of 15 - Sanitary Profiles

We look at sheets 7 and 8 together, since they represent other crucial utility information within the set. These sheets provide important utility information, such as water and wastewater connections. They also show the storm sewer system to ensure no conflicts between the various utilities during construction. We start with the existing manhole SMH-1 designated on the previous sheet and design other sanitary manholes on these sheets.

The distance between the first manhole SMH-1 at station 24 +/- to the final manhole, SMH-3 at station 5+98 at the cul-de-sac is approximately 574 feet. The second manhole, SMH-2, will be located approximately halfway at station 294 +/-. The last manhole, SMH-3, will be set at 304 feet from the manhole SMH-2.

Each manhole has an invert elevation of 0.45% above the previous manhole's invert elevation. This 0.45% grade is the minimum grade to allow gravity flow from the structure SMH3 to SMH-1. The rim elevations are to match the proposed ground elevation.

Different agencies require anywhere between 0.40% to 0.50% grade to meet the minimum velocity of two feet per second for full flow conditions. Lower grade can be used for larger pipe diameter, or the pipe can be laid flatter. It is essential to mention that the pipe's material and the diameter will impact the design grade since the manning "n" value is different for different materials.

The waterline is connected to the existing eight-inch water main located on Walden Woods Drive's east side. This connection will provide a six-inch polyvinyl chloride (PVC) water line to the front of the property. An eight-inch by six-inch tee and a gate valve is used to make the connection. The utility purveyor typically performs this type of connection. The contractor will temporarily install a cap until the proposed waterline to the subdivision can be extended.

Later, the cap will be removed. The six-inch waterline will be extended on the north side of the proposed Prairie Meadow Lane to maintain a minimum of ten feet from the proposed sanitary line located on the lane's south side and center.

Waterline and sanitary line laterals can be installed from the mainline to each lot, as shown on Sheet 7 of 15. Also, water meters will be constructed at the end of each lot.

Traffic signs are also indicated since this is a small project and there is enough space on this sheet. Let us review some of the information provided on Sheets 7 and 8 of 15.

Utility Notes:

These notes provide some minimum information regarding standards and agencies having jurisdiction.

Standard Water/Sewer Separation Statement:

These notes provide minimum information regarding horizontal and vertical separation requirements between water, sanitary, and storm sewer lines.

Manhole Adjustment Detail:

This detail provides the Sanitary manhole frame adjustment to match the proposed grade changes. An adjustment may be necessary due to surrounding grade changes, asphalt overlay, roadway improvements, etc.

Manhole Frame and Cover Detail:

This detail is used for a new manhole construction within the AHJ. Each utility district has its information. Although there may not be significant differences between most of those details for a particular use, paying attention to the nuisances is required and can save time and money.

Sign Tabulation Schedule:

This is a simple table indicating what signs are required for the proposed lane and cul-de-sac. All these signs must meet the *Manual on Uniform Traffic Control Devices (MUTCD)*. Depending on the location of the project and the AHJ, more elaborate and detailed Traffic Control (TC) plans, also called the Maintenance of Traffic (MOT) plans, should be prepared by the contractor, and submitted to the AHJ for review and approval before construction commencement.

Note: Additional sanitary sewer, potable water, and general details are presented on sheets 10 and 11 of 15.

The most important thing to be mindful of in utility design is conflicts. Any permit with digging should include an 811 Call Before You Dig label in Florida, but this is especially important for utilities. This number may be different in other states or regions. It is number 411 in the State of Washington. When looking at proposed utilities, it is crucial to consider both the horizontal and vertical position of all utilities along the property's frontage to ensure no proposed installations occupy the same space. If possible, a plan sheet should show this information for all utilities to make it readily apparent where utilities should be installed relative to each other.

Jurisdictions are increasingly demanding or encouraging the undergrounding of electrical and communication connections. There are also rules for horizontal space between close water and sewer lines (generally, a ten-feet minimum at least) that are important to keep in mind. Some utilities may require additional consideration. That can have a high cost, and it is best to investigate if that is necessary for the project early. If frontage improvements are needed, what is the impact on these installations?

Before utility design, in many cases, one will need to pothole or request records from the city, county, or utility for the existing installations to determine where you can tie into the system.

For more about potholing, refer to the ROW construction permit type in Chapter 5. Any utility installation in the ROW that blocks part of or is immediately adjacent to a sidewalk or road will need a traffic control plan.

7) Sheet 9 of 15 - Landscape Plan, Notes, and Specifications

Landscape and irrigation requirements can significantly vary from agency to agency and based on the project's scope and size. Some may require a more extensive tree replacement ratio, and others may not require any. A landscape architect designs this sheet and should be signed and dated before submission to AHJ.

Detailed construction plans to include general notes to the contractor must be provided on protecting the existing trees from damages during the construction. This drawing shows the number of existing trees removed and those to remain. For those trees to be removed, justification and compensation must be proposed for the AHJ approval.

8) Sheet 10 of 15 - General Details

Let us review some of the water details provided on this sheet.

a. Typical Service Connection

The utility purveyor provides this detail to help the utility contractor connect the service line to the main water line. In this case, a six-inch lateral connects to an eight-inch water main.

b. Valve Jacket Collar Detail

The utility purveyor provides this detail to protect the proposed valve from damage. Note 4 indicates a requirement to construct concrete valve jacket collar overall valves within the unpaved area.

c. Water Meter Connection Detail

The utility purveyor provides this detail to show how and where the water meter should be installed. As shown in this detail and indicated on sheet 7 of 15, water meters are located on private property just behind the ROW line. A city employee can access the meter from the sidewalk or the city ROW to read or repair the meter. By the way, some agencies do not allow private contractors to make connections or install the meter. These tasks are only allowed to be performed by their staff. Some other agencies allow it with the presence of their inspector.

This detail, again, is specific to the AHJ. Sometimes may even vary in size and color. Contacting the AHJ and ensuring their details are being proposed is imperative to save time and money.

The above details all pertain to the potable water main and service connections. Now, let us review some of the general details provided on this sheet.

Type "D" Curb Detail

The AHJ provided this information to be used for internal residential streets. In this case, it is used for Prairie Meadows Lane.

b. Type "F" Drop Curb Detail

This detail is used where driveways are located to facilitate the driving surface. There must be a smooth transition from the D curb to the F curb.

c. Type "F" Curb Detail

This detail is provided on collector streets such as Walden Woods Drive.

d. Typical Cul-de-sac Detail

The AHJ provides this detail to ensure the proper dimensions are used in constructing a cul-de-sac. This detail is different for different agencies, and it has to do with the size of the fire truck their fire department uses.

The landscape island is optional and can be a gravel bed or paved area. Three magnolia trees were proposed within the island to reflect the subdivision's name for this project.

e. Sidewalk Details

Some agencies have sidewalk details, and others do not. Regardless, most agencies have requirements concerning the dimensions and materials of the sidewalk in their municipal codes that need to be followed. This detail indicates a six-inch concrete, five-foot-wide sidewalk with control joints and expansion joints specifications. It also requires welded wire mesh reinforcement that is not required by many other jurisdictions.

f.   Underdrain Detail

This detail is designed and prepared by the engineer to filter and percolate additional stormwater through this system. If the pond bottom area was large enough, there was no need for this system. It is needed to compensate for the shortage of percolation capability through the proposed pond's bottom area. Surface water management districts or agencies should have the requirements and specifications for this type of system.

g.   Clear Visibility Triangle

This detail ensures that no tall structures or trees are installed within the clear visibility triangle. The purpose is for a driver entering the adjacent road must have a clear view above their sight to make a safe entry.

h.   Typical Sign Elevation

In this case, a stop sign must meet the *Manual on the Uniform Traffic Control Devices (MUTCD)* requirements or other applicable governmental agencies' rules and regulations. Some private organizations propose signs of different color materials or sizes but still must follow certain local agencies' requirements and obtain approval before installation.

i. Typical Pavement Section

This section is an essential element in a project. It directs the contractor on what dimensions, materials, and specifications to use to build the roadway, which is crucial for any subdivision. Some agencies may provide a minimum structural number for the engineer to design and meet. Other agencies may offer the exact typical section with each layer defined to be used by the engineer and their contractor. Regardless, adhering to these requirements is paramount and can avoid costly correction notices from the inspectors.

9) Sheet 11 of 15 - Sanitary Details and Notes

The intent is to provide construction details and specifications for the sanitary sewer line construction. Some of these details and specs are specific to the jurisdiction and supplied by the

AHJ. The engineer designing and preparing the sanitary sewer construction plans provides some others.

Let us review some of the details and information provided on this sheet.

1.   Typical Service Lateral Detail

This detail provides single and double service connections from the main sanitary sewer line to the facility (in this case, a single-family residence). The AHJ supplies this and it includes location, dimensions, cover, and specification for the required materials. It is important to note that the construction needs to be scheduled for inspection by the AHJ field inspector before covering the excavated area.

2.   Typical Cleanout Detail

A cleanout is necessary to maintain and clean the lateral sanitary sewer lines. It acts almost like a manhole with much less expense. Of course, a manhole is used over the main sewer lines. The required maximum distance between cleanouts is eight feet. However, some permitting authorities may impose more stringent criteria (less than eighty feet).

3.   Minimum Water/Sanitary Sewer Clearances

This detail provides the required minimum horizontal and vertical clearances between water and sanitary and storm sewer lines. These are minimum standards, and some permitting agencies may require more than just the minimum. It also depends on the site conditions, such as soil types, depth, water table, etc., to increase or provide other protection devices such as casing.

4.   Standard Shallow Manhole

This detail shows a precast concrete shallow manhole. It provides the necessary dimensions and specifications for a concrete manufacturing company to construct it. It also varies for variance permitting agencies. Some agencies may have specific requirements and instructions for the manhole top, such as the stamped agency's name or logo.

5.   Standard Manhole

This detail is like the shallow precast concrete sanitary manhole, except that it has a barrel. The length of the barrel can vary to a limit before impacting the structural design of the manhole. Regardless of the manhole height, many precast manufacturing companies in the US design and cast these structures to fit just any field condition.

6. Drop Manhole

Drop manhole is required by the AHJ when an inflow drop of two feet or more inside a manhole. The purpose is to avoid scouring and splashing inside a manhole. Remember that these manholes are subject to routine inspection and maintenance entering the manhole.

7. Sectional Invert Plan

This detail provides the required or preferred construction angle between inflow and outflow pipes to facilitate the effluent transition.

8. Specifications

This section provides both construction and materials specifications to the contractor performing the task. It gives some standard requirements and specifications for pipes and fitting materials, line cutting and connections at the structures (manholes), permit agencies' requirements, testing and televising for inspection, etc. Many of these specifications also refer to the details provided on this sheet to clarify the intent.

10) Sheet 12 of 15 - Best Management Practices Plan

This sheet is called the temporary erosion and sediment control (TESC) plan. This plan aims to ensure that stormwater runoff does not create erosion and sediment discharge to the neighboring properties and where the project is adjacent to a water body (i.e., creek, river, wetland). Polluted and untreated discharge to natural water bodies can adversely impact the species living in those water bodies by increasing turbidity and water pollution. Essentially, any untreated stormwater runoff must be controlled and kept on site. Before discharging to the existing water body the permitted construction site or any allowed overflow must be treated through various means and methods.

As part of the National Pollutant Discharge Elimination System (NPDES) Notice of Intent (NOI) General Construction Permit, all projects with a disturbed area of one acre or more must have prepared a TESC plan and submitted to NPDES for NOI approval.

Most AHJ in Washington State requires a certified erosion and sediment control lead (CESCL) to have input during the design and construction and inspect the temporary erosion and sediment control (TESC) plans. The CESCL is a certified professional trained by the Department of Ecology (DOE) or other qualified agencies in Washington State to routinely inspect the site during construction to ensure that the site's temporary BMP is always maintained.

The DOE also provides training programs for those who like to be trained and certified as a CESCL. They keep a list of CESCL inspectors for those interested in hiring a CESCL.

Other states may use the Department of Environmental Protection Agency (EPA) Certified Professional in Erosion and Sediment Control (CPESC), which embraces surface erosion and sediment control science. This practice also specializes in the study and subsequent reduction of the adverse effects of environmental pollutants, whether natural or manufactured, as it relates to the soil, water, and air. The CPESC is the only professional certification recognized by the EPA whose professionals are designated as qualified to prepare Stormwater Pollution Prevention Plans (SWPPPs) and reports.

Each number on this sheet corresponds to a detail provided on the next sheet. Implementing each element performs a specific task to accomplish a certain degree of control to avoid unwanted discharge. Therefore, each component can only be applied to a particular location, as shown here.

11) Sheet 13 of 15 - Best Management Practices Details

Each detail on this sheet has an assigned number to indicate where it should be installed on the previous sheet and implemented in the construction field. They include sixteen specific details that some of those details relate to this project, but not all do. Details numbers 1, 8, 10, 13, and 14A are used for this project, and their locations are shown on the last sheet, Best Management Practices Plan.

Let us review some of the details provided on this sheet.

1. Typical Retention/Detention Pond Section

This detail, combined with detail number 3, is a simple section that indicates the pond's side slope plus the sodding requirement. The reader may also want to review the definitions and differences between a retention and detention pond. Please note that most of the stormwater runoff storage facilities are detention facilities.

2. Temporary Sediment Sump

This temporary sediment sump (TSS) is another simple detail showing a TSS during the construction. It can get a lot more complicated, and it may require a detailed computation and design to handle any sediment trap during construction for much larger projects.

3.   Grass Slopes

Side Slopes with grass are another simple detail requiring either sodding or seed implementation based on side steepness.

4.   Typical Swale Section

This detail provides side slopes and minimum dimensions for a typical swale and minimum limit requiring sodding.

5.   Silt Barrier at the Connection of Swale to Existing Swale

The silt barrier stops silt and other construction debris from coming through the proposed swale by providing silt barriers, hay bales, or a combination of both, as shown. This detail aims to block and capture illicit discharges that can be removed manually. It ensures that the foreign materials will not reach a water body downstream of the swale.

6.   Silt Barrier at the Connection of Storm Pipe to Existing Swale

To block and capture illicit discharges such as silt and other construction debris coming through the pipe, silt fence, hay bale, or a combination of both is required, as shown in this detail. This detail is the same as the last detail, except it is located in the proposed storm pipe entering a ditch. It ensures that the foreign materials will not reach a water body downstream.

7.   Silt Barrier at the Connection of Storm Pipe to Existing Lake

This detail is the same as the last detail, number 6. The exception is that the shape of the barrier is configured to accommodate the lake's geometry.

8.   Stabilized Construction Exit

Trucks and other vehicles accessing a construction site typically have dirty tires that can create track-out on clean roadway surfaces if not cleaned. Besides tracking dirt and other materials to the streets, those materials can create hazardous driving conditions, particularly during the wet season. This detail requires all vehicles leaving the site to wash all debris from their tires before accessing the public ROW. Some contractors prefer to use other more sophisticated wheel washes in the market, especially for larger projects.

9. Sod Along Curb and Around Inlet

This detail ensures proper sodding vs. hydroseeding behind the back of the curb and around the inlet, assuming that either method is acceptable by the AHJ. Some agencies or projects may not allow hydroseeding due to the weather or other conditions, and sodding must be implemented.

10. Haybale Protection Around Inlets or Similar Structures

These details indicate the hay bale's proper location around different inlet types, such as gutter inlets and ditch bottom inlets.

11. Staked Silt Barrier or Silt Fence Protection Around Ditch Bottom Inlets

This detail provides directions to locate proper hay bales and filter fabric to protect a ditch bottom inlet.

12. Underground Pipe Crossing

This detail is for constructing an under-ditch pipe (i.e., water, sanitary sewer). The designer and the contractor need to explore the dewatering or other techniques before being implemented in a wet condition. This detail is a lot easier to implement if the ditch is dry.

13. Hay Bale Detail

This detail provides directions for constructing hay bales on slopes.

14. Typical Silt Fence

This detail provides directions for constructing a typical silt fence per the Florida Department of Transportation (FDOT) when designing and preparing these construction plans. There could be some changes to this detail, and the reader is encouraged to contact the FDOT website for more up-to-date information.

15. Silt Barrier along Wetlands

This general detail provides directions for constructing a typical silt barrier to separate uplands (construction area) from the wetlands. The buffer dimensions vary per the local, state, and other environmental rules and regulations. The actual detail may also vary per the permitting agency's requirements and sometimes based on the importance or criticality of the wetlands. For example, the condition may be to provide double lines of silt barrier to ensure additional wetlands protection.

16. Established Earth Berm

This general detail provides directions for constructing a typical stabilized temporary earth berm downhill of an uphill construction area. As the note explains, the earth berm will be removed, and the disturbed area will be restored after construction.

12) Sheet 14 of 15 - Pre-Development Drainage Plan

The primary purpose is to provide pre-development drainage data to facilitate drainage computation and report. It is not a construction plan sheet and does not need to be included in construction plans offered to contractors for bidding, except for information only. However, it is an integral part of the drainage report submitted to permitting agencies reviewing and approving stormwater management design.

Let us review some of the information provided on this sheet.

It is essential to mention here that the pre-development drainage plan can utilize the actual survey sheet with all existing features. Basin drainage boundary should be delineated on this sheet. The area within the basin drainage boundary is the total pre-development basin area. Since the entire site is relatively flat and the stormwater runoff sheet flows toward the northeast corner, there is no need further to divide the drainage basin boundary into smaller sub-basin areas. Of course, this is not always the case and depends on the topography and existing facilities (if any). The flow length, average slope, and time of concentration are also indicated.

Time of Concentration Computation $Tc = Tt1 + Tt2 + ...$

Tt is the travel time, the time it takes water to travel from one location to another in a watershed.

In this case, a water drop travels from the watershed's northeast corner to the southeast corner. Therefore, there is only one travel time, the same as the time of concentration.

$Tt = HL / 60 V$

Where:

T_t = Travel time (minute) HL = Flow length (feet)
V = Average velocity (feet/second)
60 = conversion factor from seconds to minutes.

HL = 720 feet

Ave. slope = (145.1' - 142.0') / 720 feet = /720 = 0.0043 ft./ft.

Based on the above calculated average slope and the type of ground coverage and using the graph presented on the following page.

V = 0.217 feet/second

(Assuming the ground cover is somewhere between forest with heavy ground litter and hay meadows - minimum tillage cultivation, contour or strip-cropped, and woodland).

Therefore, T_c = T_t = 720 feet / (60 x 0.217 feet/second) = 55.31 minutes

The above calculated T_c means that it takes a raindrop approximately 55.31 minutes to travel from the northwest corner to reach the southeast corner to discharge out of the site. In theory, any other raindrop reaches the northwest corner or some other final point to bounce out of the site in a shorter time.

13) Sheet 15 of 15 - Post-Development Drainage Plan

The primary purpose is to provide post-development drainage data to facilitate drainage computation and reporting. It is not a construction plan sheet and does not need to be included in construction plans offered to contractors for bidding, except for information only. However, it is included in the drainage report submitted to permitting agencies reviewing and approving stormwater management design.

In addition to delineating the basin drainage boundaries, it is essential to subdivide the entire drainage area into smaller sub-basin drainage boundaries to facilitate computations. Since the existing site is relatively flat and the storm runoff sheet flows from the east to the west (lower elevations), it makes engineering and economic sense to collect post-construction stormwater runoff on the east (lower) side. Therefore, we proposed two shallows (dry) detention ponds on the east side of the project as discussed in the previous Paving, Grading, and Drainage Plan Sheet 5 of 16 in this chapter.

Let us review some of the information provided on this sheet.

Drainage Area #1:

This drainage area includes lots 1 to 8 plus the north half of the Prairie Meadows Lane drainage. The flow length is calculated by the storm runoff sheet flowing from lot 8 to the cul-desac area and then thru the proposed gutter, reaching the inlet S-2. The time of concentration is calculated the same way as for the pre-development drainage.

Drainage Area #2:

This drainage area includes lots 9 to 16 plus the south half of the Prairie Meadows Lane drainage. The flow length is calculated by the storm runoff sheet flowing from lot 9 to the cul-desac area and then thru the proposed gutter, reaching the inlet S-3. The time of concentration is calculated the same way as for the pre-development drainage.

Drainage Area #3 & 4:

These drainage areas include Pond 1 and Pond 2 areas, respectfully. The flow length for these two areas is very short, and the concentration-time is almost instant. A time of concentration of ten minutes is assumed for each of these two drainage areas to meet the minimum Tc requirement.

The following pages represent the Construction Plans Sheet 1 through 15.

**Final Construction Plans
(Sheets 1 to 15 of 15)**

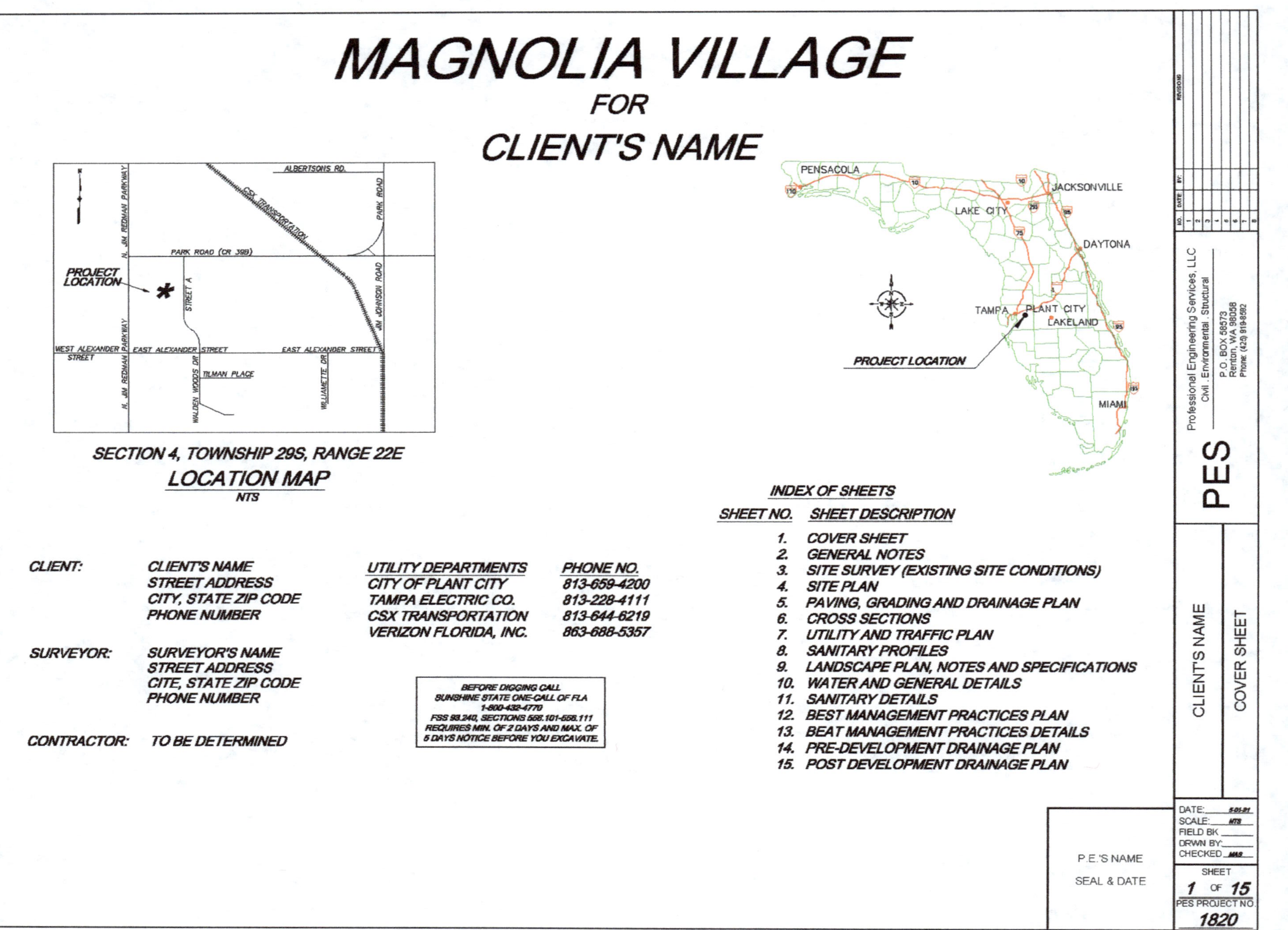

ALI SHASTI, P.E.

MAGNOLIA VILLAGE
FOR
CLIENT'S NAME

ALBERTSONS RD.
CSX TRANSPORTATION
PARK ROAD
N. JIM REDMAN PARKWAY
PARK ROAD (CR 39B)
PROJECT LOCATION
STREET A
JIM JOHNSON ROAD
WEST ALEXANDER STREET
EAST ALEXANDER STREET
EAST ALEXANDER STREET
N. JIM REDMAN PARKWAY
WALDEN WOODS DR
TILMAN PLACE
WILLIAMETTE DR

SECTION 4, TOWNSHIP 29S, RANGE 22E
LOCATION MAP
NTS

PENSACOLA
JACKSONVILLE
LAKE CITY
DAYTONA
TAMPA
PLANT CITY
LAKELAND
MIAMI
PROJECT LOCATION

CLIENT: CLIENT'S NAME
                STREET ADDRESS
                CITY, STATE ZIP CODE
                PHONE NUMBER

SURVEYOR: SURVEYOR'S NAME
                STREET ADDRESS
                CITE, STATE ZIP CODE
                PHONE NUMBER

CONTRACTOR: TO BE DETERMINED

UTILITY DEPARTMENTS   PHONE NO.
CITY OF PLANT CITY   813-659-4200
TAMPA ELECTRIC CO.   813-228-4111
CSX TRANSPORTATION   813-644-6219
VERIZON FLORIDA, INC.   863-688-5357

BEFORE DIGGING CALL
SUNSHINE STATE ONE-CALL OF FLA
1-800-432-4770
FSS 98.240, SECTIONS 588.101-588.111
REQUIRES MIN. OF 2 DAYS AND MAX. OF
5 DAYS NOTICE BEFORE YOU EXCAVATE.

INDEX OF SHEETS
SHEET NO.   SHEET DESCRIPTION
1.   COVER SHEET
2.   GENERAL NOTES
3.   SITE SURVEY (EXISTING SITE CONDITIONS)
4.   SITE PLAN
5.   PAVING, GRADING AND DRAINAGE PLAN
6.   CROSS SECTIONS
7.   UTILITY AND TRAFFIC PLAN
8.   SANITARY PROFILES
9.   LANDSCAPE PLAN, NOTES AND SPECIFICATIONS
10. WATER AND GENERAL DETAILS
11. SANITARY DETAILS
12. BEST MANAGEMENT PRACTICES PLAN
13. BEAT MANAGEMENT PRACTICES DETAILS
14. PRE-DEVELOPMENT DRAINAGE PLAN
15. POST DEVELOPMENT DRAINAGE PLAN

PES
Professional Engineering Services, LLC
Civil . Environmental . Structural
P.O. BOX 58573
Renton, WA 98058
Phone (425) 919-8592

CLIENT'S NAME
COVER SHEET

DATE: 5-01-21
SCALE: NTS
FIELD BK
DRWN BY:
CHECKED MAS

P.E.'S NAME
SEAL & DATE

SHEET
1 OF 15
PES PROJECT NO.
1820

142

GENERAL NOTES AND SPECIFICATIONS

1. ALL CONSTRUCTION METHODS, WORKMANSHIP, MATERIALS, AND TESTING, SHALL CONFORM TO THE "FLORIDA DEPARTMENT OF TRANSPORTATION (FDOT)'S "STANDARD SPECIFICATION FOR ROAD AND BRIDGE CONSTRUCTION, " ITS CURRENT SUPPLEMENTS AND STANDARD INDEX DRAWINGS. THEY ALSO MUST MEET THE CITY AND ANY OTHER APPLICABLE AGENCY'S RULES AND REGULATIONS UNLESS OTHERWISE NOTED ON THE PLANS.

2. SURVEY LAYOUT AND TESTING SHALL BE DONE BY THE CONTRACTOR.

3. ALL WORK WITHIN THE RIGHT-OF-WAY SHALL BE COORDINATED WITH THE CITY. ALL DISTURBED AREAS SHALL BE RESTORED TO THE ORIGINAL CONDITION OR BETTER PER CITY STANDARDS.

4. LOCATIONS, ELEVATIONS, AND DIMENSIONS OF EXISTING UTILITIES, STRUCTURES, AND OTHER FEATURES ARE SHOWN ACCORDING TO THE BEST INFORMATION AVAILABLE AT THE TIME OF PREPARATION OF THESE PLANS. THE CONTRACTOR SHALL VERIFY THE LOCATIONS, ELEVATIONS, AND DIMENSIONS OF EXISTING UTILITIES, STRUCTURES, AND OTHER FEATURES AFFECTING THIS WORK BEFORE THE CONSTRUCTION COMMENCEMENT.

5. THE CONTRACTOR SHALL USE EXTREME CAUTION IN AREAS OF BURIED UTILITIES AND SHALL PROVIDE AT LEAST 48 HOURS' NOTICE TO THE UTILITY COMPANIES BEFORE CONSTRUCTION TO OBTAIN FIELD LOCATIONS OF EXISTING UNDERGROUND UTILITIES. THE CONTRACTOR SHALL PROTECT AND MAINTAIN THE OPERATION OF EXISTING UTILITIES DURING CONSTRUCTION.

6. THE CONTRACTOR SHALL CHECK PLANS FOR CONFLICTS AND DISCREPANCIES BEFORE CONSTRUCTION. THE CONTRACTOR SHALL NOTIFY THE ENGINEER OF ANY CONFLICTS BEFORE PERFORMING ANY WORK IN THE AFFECTED AREA.

7. THE CONTRACTOR MUST NOTIFY THE ENGINEER AS TO WHERE EXISTING UTILITY SERVICES INTERFERE WITH THE PROPOSED CONSTRUCTION.

8. THE CONTRACTOR IS RESPONSIBLE FOR REPAIRING ANY DAMAGE TO EXISTING FACILITIES, ABOVE OR BELOW GROUND, THAT MAY OCCUR AS A RESULT OF THE WORK PERFORMED BY THE CONTRACTOR.

9. ALL UNDERGROUND UTILITIES MUST BE IN PLACE AND TESTED OR INSPECTED PRIOR TO BASE AND SURFACE CONSTRUCTION.

10. UTILITY INSTALLATION SHALL CONFORM TO THE REQUIREMENTS OF THE APPLICABLE FEDERAL, STATE, AND LOCAL REGULATIONS, INCLUDING BUT NOT LIMITED TO THE FLORIDA DEPARTMENT OF ENVIRONMENTAL PROTECTION (FDEP), DEPARTMENT OF HEALTH (DOH), THE OCCUPATIONAL SAFETY AND HEALTH ADMINISTRATION (OSHA) STANDARD AND THE CITY.

11. IT IS THE CONTRACTOR'S RESPONSIBILITY TO BECOME FAMILIAR WITH THE PERMIT AND INSPECTION REQUIREMENTS SPECIFIED BY THE VARIOUS GOVERNMENTAL AGENCIES AND THE ENGINEER. THE CONTRACTOR SHALL OBTAIN ALL NECESSARY PERMITS PRIOR TO CONSTRUCTION AND SCHEDULE INSPECTIONS ACCORDING TO AGENCY INSTRUCTIONS. ALL WORK PERFORMED SHALL COMPLY WITH THE REGULATIONS AND ORDINANCES OF THE VARIOUS GOVERNMENTAL AGENCIES HAVING JURISDICTION OVER THE WORK.

12. AT LEAST THREE (3) WORKING DAYS BEFORE THE CONSTRUCTION COMMENCEMENT, THE CONTRACTOR SHALL NOTIFY THE ENGINEER AND APPROPRIATE AGENCIES AND SUPPLY THEM WITH THE CONTRACTOR'S NAME, STARTING DATE, PROJECTED CONSTRUCTION SCHEDULE, ALL REQUIRED SHOP DRAWINGS, AND OTHER INFORMATION AS REQUIRED. ANY WORK PERFORMED PRIOR TO PROPER NOTIFICATIONS MAY BE SUBJECT TO REMOVAL AND REPLACEMENT AT THE CONTRACTOR'S EXPENSE.

13. THE SITE SHALL BE INITIALLY GRADED SUCH THAT NO OFFSITE AREA WILL BE ADVERSELY AFFECTED BY STORMWATER RUNOFF. RUNOFF TO THE RIGHT-OF-WAY SHALL ALSO BE MINIMIZED. ALL CONSTRUCTION AND DISTURBED AREAS SHALL BE PROTECTED FROM EROSION CAUSED BY STORMWATER RUNOFF BY UTILIZING THE EROSION AND SEDIMENT CONTROL MEASURES INDICATED ON THE BEST MANAGEMENT PRACTICES PLANS AND DETAILS.

14. UPON COMPLETION OF CONSTRUCTION ACTIVITIES, THE STORMWATER MANAGEMENT SYSTEM SHALL BE CLEANED OF ALL DEBRIS, SILTS, LIMEROCK, ETC.

15. THE CONTRACTOR SHALL VERIFY MINIMUM SEPARATION BETWEEN WATER AND SANITARY SEWER SYSTEMS (18" MINIMUM VERTICAL AND 10' MINIMUM HORIZONTAL SEPARATIONS). UPON COMPLETION OF CONSTRUCTION, THE CONTRACTOR SHALL PROVIDE WRITTEN VERIFICATION DOCUMENTING THE MINIMUM SEPARATION AS CONSTRUCTED.

16. TYPICAL COVER OVER UTILITIES SHALL BE 36".

17. DURING THE CONSTRUCTION AND MAINTENANCE OF THIS PROJECT, ALL SAFETY REGULATIONS ARE TO BE ENFORCED. THE CONTRACTOR OR HIS REPRESENTATIVE IS RESPONSIBLE FOR THE CONTROL AND SAFETY OF THEIR PERSONNEL AND THE TRAVELING PUBLIC.

18. THE CONTRACTOR SHALL COMPLY WITH ALL FEDERAL AND STATE OCCUPATIONAL SAFETY, HEALTH ACT (OSHA) STANDARDS, ANY OTHER RULES, AND REGULATIONS APPLICABLE TO THE CONSTRUCTION OR MAINTENANCE ACTIVITIES IN THE STATE OF FLORIDA. THE CONTRACTOR SHALL ALSO COMPLY WITH CHAPTER 442, FLORIDA STATUTES (TOXIC SUBSTANCES IN THE WORKPLACE), AND ANY CITY OR ANY OTHER AGENCY'S RULES AND REGULATIONS REGARDING SAFETY.

19. ALL TRENCHING AND EXCAVATION OPERATIONS MUST CONFORM TO THE REQUIREMENTS STATED IN THE NEW "TRENCH SAFETY ACT" INCORPORATED INTO OSHA STANDARDS 29 CFR 1926 SUBPART P FINAL RULE AND THE STATE OF FLORIDA TRENCH SAFETY ACT. THE CONTRACTOR SHALL RECOGNIZE THE OSHA SAFETY STANDARDS, AGREE TO ABIDE BY THEM, AND IDENTIFY THE COST TO COMPLY.

20. FURTHER CONTRACTOR OPERATIONS SHALL COMPLY WITH 29CFR 1910.252 AND NFPA 51B FOR CUTTING AND WELDING PROCEDURES.

21. ALL PERSONS ON (CITY OR STATE) PROPERTY AND IN AREAS WHERE THE NOISE LEVEL EXCEEDS 85DB, MUST WEAR HEARING PROTECTION THAT COMPLIES WITH ANSI S3.19-74 (EARMUFFS OR APPROVED EARPLUGS). THIS INCLUDES AREAS WHERE NOISY EQUIPMENT IS IN USE (I.E., JACKHAMMERS, ELECTRIC OR AIR DRILLS, HEAVY EQUIPMENT WITH OPEN CABS, PIPE CUTTING SAWS, ETC.) AND IN THE PLANT WHERE POSTED.

22. ALL WORK CONDUCTED IN AN ELEVATED POSITION SHALL COMPLY WITH 29CFR 1910.269.

23. ALL DESIGN APPLICATION, INSTALLATION, MAINTENANCE, AND REMOVAL OF ALL TRAFFIC CONTROL DEVICES, WARNING DEVICES, AND BARRIERS NECESSARY TO PROTECT THE PUBLIC AND WORKMEN FROM HAZARDS WITHIN THE PROJECT AREA SHALL BE PER THE MINIMUM STANDARDS AS OUTLINED IN THE CURRENT EDITION OF THE MANUAL ON UNIFORM TRAFFIC CONTROL DEVICES (MUTCD).

24. ALL TRAFFIC CONTROL MARKINGS AND DEVICES SHALL CONFORM TO THE PROVISION OUTLINED IN THE "MANUAL ON UNIFORM TRAFFIC CONTROL DEVICES" PREPARED BY THE U.S. DEPARTMENT OF TRANSPORTATION, FEDERAL HIGHWAY ADMINISTRATION.

25. ALL CONTRACTOR-OWNED OR CONTROLLED VEHICLES AND PIECES OF EQUIPMENT THAT WILL BE OPERATED ON OR WITHIN TEN (10) FEET OF THE ROADWAY SHALL BE EQUIPPED WITH A MINIMUM OF ONE AMBER 360 DEGREE CLASS I WARNING DEVICE. THIS DEVICE MUST MEET MINIMUM STANDARDS FOR UTILITY CONSTRUCTION PURPOSES, SUCH AS A MINIMUM OF 500,000 CANDLEPOWER AND VISIBLE FROM 360 DEGREES OF MOUNTING. THE WARNING DEVICE(S) MUST BE IN OPERATION AT ALL TIMES THAT A VEHICLE/EQUIPMENT IS ON THE ROADWAY OR WITHIN THE TEN (10) FEET OF RUNOFF AREA AND NOT IN A "NORMAL" TRAVEL STATUS.

26. AREA DESIGNATED BY THE OWNER SHALL BE UTILIZED FOR EQUIPMENT STORAGE AND COVER MATERIAL STAGING. MATERIALS AND SUPPLIES SHALL BE STORED ONLY IN THOSE AREAS APPROVED BY THE OWNER.

27. REGARDING THE PROJECT'S INSPECTION, SOUTHWEST FLORIDA WATER MANAGEMENT DISTRICT (SWFWMD) AND THE CITY INSPECTORS HAVE THE SAME RIGHT OF REVIEW AND INSPECTION AS THE ENGINEER OF RECORD. ALTHOUGH IT IS APPROPRIATE FOR THE ENGINEER OF RECORD TO RESOLVE QUESTIONS, DIFFICULTIES, AND DISPUTES RELATIVE TO INTERPRETATION OF CONTRACT DOCUMENTS OR CONSTRUCTION, SWFWMD AND THE CITY RESERVE THE RIGHT OF FINAL CONCURRENCE ON SUCH MATTERS.

28. THE CONTRACTOR SHALL OBTAIN THE ENGINEER AND THE CITY APPROVAL ON ALL PROPOSED FIELD CHANGES, ALTERATIONS, OR ADDITIONS TO THE APPROVED PLANS BY THE CONTRACTOR BEFORE THE IMPLEMENTATION OF SUCH CHANGES, ALTERATIONS, OR ADDITIONS.

29. THE CONTRACTOR SHALL OBTAIN A SEPARATE PERMIT FOR ANY ON-SITE BURNING (IF ALLOWED).

30. DURING CONSTRUCTION ACTIVITIES, ANY EVIDENCE OF HISTORIC RESOURCES, INCLUDING BUT NOT LIMITED TO ABORIGINAL OR HISTORIC POTTERY, PREHISTORIC STONE TOOLS, BONE OR SHELL TOOLS, HISTORIC TRASH PITS, OR HISTORIC BUILDING FOUNDATION, ARE DISCOVERED, WORK SHALL COME TO AN IMMEDIATE STOP. THE CONTRACTOR SHALL NOTIFY THE FLORIDA DEPARTMENT OF HISTORIC RESOURCES (STATE HISTORIC PRESERVATION OFFICER) AND THE CITY WITHIN TWO WORKING DAYS OF THE RESOURCES FOUND ON THE SITE.

31. ACCESSIBLE PARKING (ADA) SPACES WILL BE PROPERLY SIGNED AND STRIPED PER FLORIDA STATUTE 316, THE MANUAL ON UNIFORM TRAFFIC CONTROL DEVICES, OR ANY OTHER APPLICABLE STANDARDS.

32. ALL ON-SITE PARKING SPACES WILL BE STRIPPED AND SIGNED FOLLOWING THE MANUAL ON UNIFORM TRAFFIC CONTROL DEVICES LATEST EDITION. PARKING SPACES, DIRECTIONAL ARROWS, AND STOP BARS SHALL BE STRIPED IN WHITE. IT SHALL BE THE OWNER/DEVELOPERS RESPONSIBILITY PROPERLY TO SIGN AND STRIPE THE SITE PER APPLICABLE STANDARDS.

33. ALL CLEAR ZONE AREAS SHALL BE KEPT FREE OF ANY SIGNAGE PLANTINGS, TREES, ETC. ABOVE THREE AND ONE-HALF FEET IN HEIGHT.

34. ALL STRUCTURES, INCLUDING BUFFER WALLS, RETAINING WALLS, FENCING, SIGNAGE ETC. MAY REQUIRE SEPARATE BUILDING PERMITS.

35. NO IRRIGATION SYSTEM OR LANDSCAPING SHALL BE INSTALLED IN ANY CITY RIGHT OF WAY WITHOUT THE APPROPRIATE RIGHT OF WAY PERMIT.

36. UPON COMPLETION OF THE CONSTRUCTION, THE CONTRACTOR MUST SUPPLY CERTIFIED "AS-BUILT" TO THE ENGINEER OF RECORD TO BE SUBMITTED TO THE PERMITTING AGENCIES.

RECOMMENDED CONSTRUCTION SEQUENCE

(PROVIDE THE FOLLOWING AS REQUIRED BY THE APPROVED CONSTRUCTION PLANS AND THE PERMIT CONDITIONS)

1. HOLD THE EROSION SEDIMENTATION CONTROL (ESC) PRE-CONSTRUCTION MEETING ON SITE (OPTIONAL).

2. CONDUCT AN INITIAL EROSION AND SEDIMENTATION CONTROL SITE INSPECTION.

3. HOLD THE PRE-CONSTRUCTION MEETING AT THE CITY OR ON-SITE.

4. POST A SIGN WITH THE NAME AND PHONE NUMBER OF THE SITE'S ESC SUPERVISOR.

5. FLAG AND OR FENCE THE CLEARING LIMITS.

6. INSPECTION AND SIGN-OFF OF THE CLEARING LIMITS BY A CITY SITE INSPECTOR.

7. INSTALL CATCH BASIN PROTECTION AS REQUIRED.

8. GRADE AND INSTALL CONSTRUCTION ENTRANCE(S).

9. INSTALL PERIMETER PROTECTION (SILT FENCE, BRUSH BARRIER, ETC.).

10. MARK PROPOSED BIO-RETENTION FACILITIES, RAIN GARDENS, PERMEABLE PAVEMENT, AND INSTALLING ESC BEST MANAGEMENT PRACTICES (BMPS) TO PROTECT THEM FROM COMPACTION AND SEDIMENTATION.

11. CONSTRUCT SEDIMENT PONDS AND TRAPS.

12. INSPECTION OF ESC MEASURES BY A CITY SITE INSPECTOR BEFORE THE COMMENCEMENT OF GRADING ACTIVITY.

13. GRADE AND STABILIZE CONSTRUCTION ROADS.

14. DEMOLITION OF EXISTING STRUCTURES AND SITE FEATURES.

15. CONSTRUCT SURFACE WATER CONTROLS (INTERCEPTOR DIKES, PIPE SLOPE DRAINS, ETC.). SIMULTANEOUSLY WITH CLEARING AND GRADING FOR PROJECT DEVELOPMENT.

16. MAINTAIN EROSION CONTROL MEASURES FOLLOWING CITY STANDARDS AND MANUFACTURER'S RECOMMENDATIONS.

17. RELOCATE SURFACE WATER CONTROLS AND EROSION CONTROL MEASURES OR INSTALL NEW MEASURES. AS SITE CONDITIONS CHANGE, EROSION AND SEDIMENT CONTROL MUST ALWAYS BE PER THE CITY EROSION AND SEDIMENT CONTROL STANDARDS.

18. COVER ALL AREAS THAT WILL BE UNWORKED FOR MORE THAN SEVEN DAYS DURING THE DRY SEASON OR TWO DAYS DURING THE WET SEASON WITH STRAW, WOOD FIBER MULCH, COMPOST, PLASTIC SHEETING, OR EQUIVALENT.

19. STABILIZE ALL AREAS THAT REACH FINAL GRADE WITHIN SEVEN DAYS.

20. UPON COMPLETION OF THE PROJECT, ALL DISTURBED AREAS MUST BE STABILIZED, AND BEST MANAGEMENT PRACTICES REMOVED AS APPROPRIATE.

21. BEFORE FINAL INSPECTION APPROVAL, CONDUCT AN INFILTRATION TEST FOR ANY BIO-RETENTION OR PERMEABLE PAVEMENT INSTALLATION TO CONFIRM THAT CONSTRUCTION ACTIVITIES DID NOT IMPACT THE PERMANENT BMPS. IF NEEDED, RESTORE, OR REPLACE BIO-RETENTION OR PERMEABLE PAVEMENT BMPS TO FULLY FUNCTION UPON COMPLETION OF CONSTRUCTION ACTIVITIES.

P.E.'S NAME

SEAL & DATE

Professional Engineering Services, LLC
Civil . Environmental . Structural
P.O. BOX 58573
Renton, WA 98058
Phone (425) 919-8582

PES

CLIENT'S NAME

GENERAL NOTES

DATE: 5-01-21
SCALE: NTS
FIELD BK
DRWN BY:
CHECKED: MAS

SHEET 2 OF 15

PES PROJECT NO. 1820

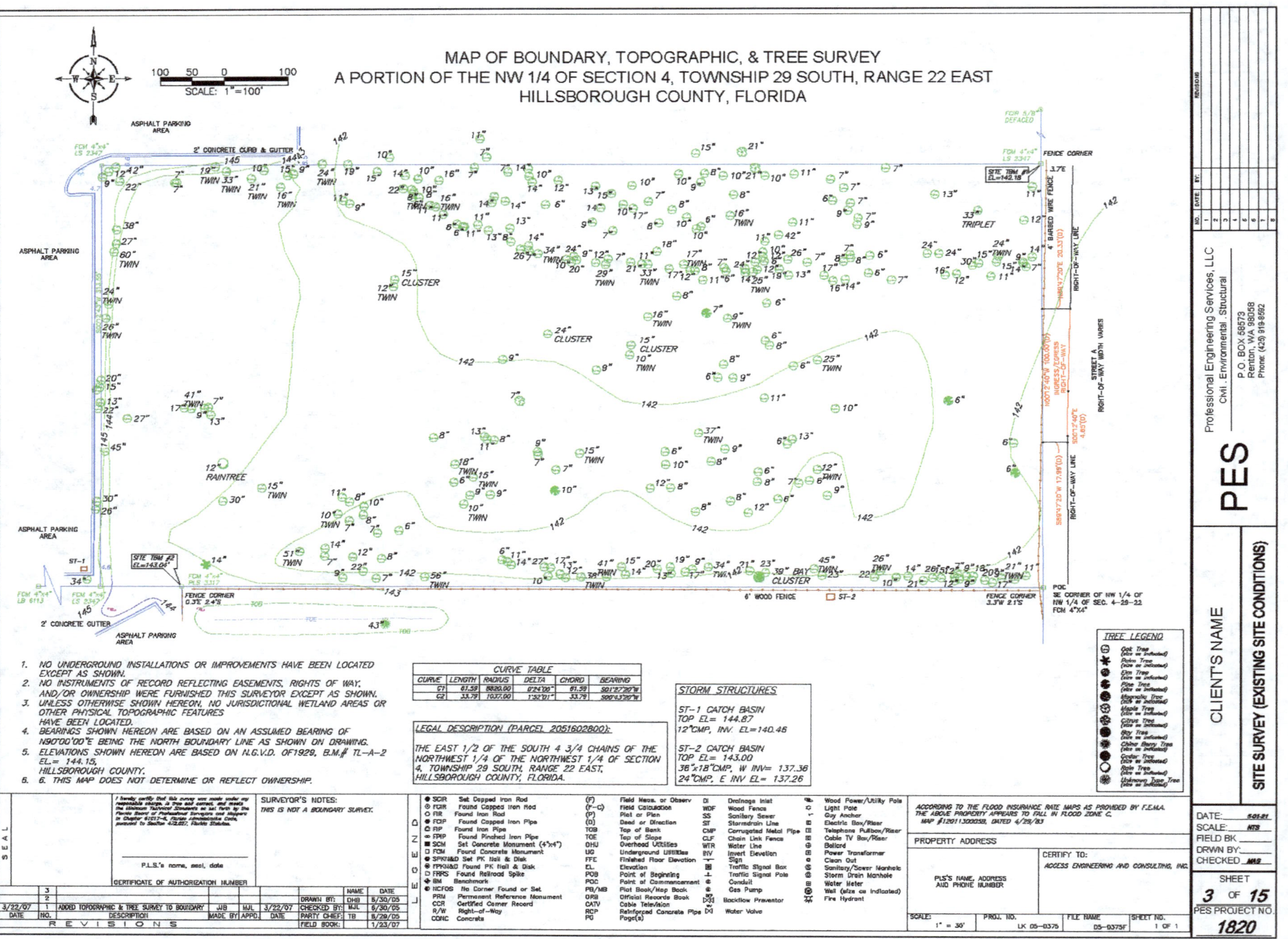

CURVE TABLE

CURVE	LENGTH	RADIUS	DELTA	CHORD	BEARING
C1	61.59	8820.00	0°24'00"	61.59	S01°27'20"W
C2	33.79	1037.00	1°52'01"	33.79	S00°43'20"W

STORM STRUCTURES

ST-1 CATCH BASIN
TOP EL= 144.87
12"CMP, INV. EL=140.46

ST-2 CATCH BASIN
TOP EL= 143.00
36"x18"CMP, W INV= 137.36
24"CMP, E INV EL= 137.26

LEGAL DESCRIPTION (PARCEL 2051602800):

THE EAST 1/2 OF THE SOUTH 4 3/4 CHAINS OF THE NORTHWEST 1/4 OF THE NORTHWEST 1/4 OF SECTION 4, TOWNSHIP 29 SOUTH, RANGE 22 EAST, HILLSBOROUGH COUNTY, FLORIDA.

1. NO UNDERGROUND INSTALLATIONS OR IMPROVEMENTS HAVE BEEN LOCATED EXCEPT AS SHOWN.
2. NO INSTRUMENTS OF RECORD REFLECTING EASEMENTS, RIGHTS OF WAY, AND/OR OWNERSHIP WERE FURNISHED THIS SURVEYOR EXCEPT AS SHOWN.
3. UNLESS OTHERWISE SHOWN HEREON, NO JURISDICTIONAL WETLAND AREAS OR OTHER PHYSICAL TOPOGRAPHIC FEATURES HAVE BEEN LOCATED.
4. BEARINGS SHOWN HEREON ARE BASED ON AN ASSUMED BEARING OF N90°00'00"E BEING THE NORTH BOUNDARY LINE AS SHOWN ON DRAWING.
5. ELEVATIONS SHOWN HEREON ARE BASED ON N.G.V.D. OF 1929, B.M.# TL-A-2 EL.= 144.15, HILLSBOROUGH COUNTY.
6. 6. THIS MAP DOES NOT DETERMINE OR REFLECT OWNERSHIP.

SURVEYOR'S NOTES:
THIS IS NOT A BOUNDARY SURVEY.

CERTIFICATE OF AUTHORIZATION NUMBER
P.L.S.'s name, seal, date

			Set Capped Iron Rod		
SCIR	Set Capped Iron Rod	(F)	Field Meas. or Observ	DI	Drainage Inlet
FCIR	Found Capped Iron Rod	(F-C)	Field Calculation	WDF	Wood Fence
FIR	Found Iron Rod	(P)	Plat or Plan	SS	Sanitary Sewer
FCIP	Found Capped Iron Pipe	(D)	Deed or Direction	ST	Stormdrain Line
FIP	Found Iron Pipe	TOB	Top of Bank	CMP	Corrugated Metal Pipe
FPIP	Found Pinched Iron Pipe	TOE	Top of Slope	CLF	Chain Link Fence
SCM	Set Concrete Monument (4"x4")	OHU	Overhead Utilities	WTR	Water Line
FCM	Found Concrete Monument	UG	Underground Utilities	RIV	Invert Elevation
SPKN&D	Set PK Nail & Disk	FFE	Finished Floor Elevation		Sign
FPKN&D	Found PK Nail & Disk	EL	Elevation		Traffic Signal Box
FRRS	Found Railroad Spike	POB	Point of Beginning		Traffic Signal Pole
BM	Benchmark	POC	Point of Commencement		Conduit
NCFOS	No Corner Found or Set	PB/MB	Plat Book/Map Book		Gas Pump
PRM	Permanent Reference Monument	ORB	Official Records Book		Backflow Preventor
CCR	Certified Corner Record	CATV	Cable Television		Fire Hydrant
R/W	Right-of-Way	RCP	Reinforced Concrete Pipe		Water Valve
CONC.	Concrete	PG	Page(s)		

Wood Power/Utility Pole · Light Pole · Guy Anchor · Electric Box/Riser · Telephone Pullbox/Riser · Cable TV Box/Riser · Bollard · Power Transformer · Clean Out · Sanitary/Sewer Manhole · Storm Drain Manhole · Water Meter · Well (size as indicated)

TREE LEGEND
Oak Tree (size as indicated)
Palm Tree (size as indicated)
Elm Tree (size as indicated)
Pine Tree (size as indicated)
Magnolia Tree (size as indicated)
Maple Tree (size as indicated)
Citrus Tree (size as indicated)
Bay Tree (size as indicated)
China Berry Tree (size as indicated)
Cedar Tree (size as indicated)
Rain Tree (size as indicated)
Unknown Type Tree (size as indicated)

ACCORDING TO THE FLOOD INSURANCE RATE MAPS AS PROVIDED BY F.E.M.A. THE ABOVE PROPERTY APPEARS TO FALL IN FLOOD ZONE C. MAP #12011300058, DATED 4/25/83

PROPERTY ADDRESS

CERTIFY TO:
ACCESS ENGINEERING AND CONSULTING, INC.

PLS'S NAME, ADDRESS AND PHONE NUMBER

DATE:	5-01-21
SCALE:	NTS
FIELD BK	
DRWN BY:	
CHECKED	MAS

SCALE:	PROJ. NO.	FILE NAME	SHEET NO.
1" = 30'	LK 05-0375	05-0375F	1 OF 1

SHEET 3 OF 15
PES PROJECT NO. 1820

CLIENT'S NAME
SITE SURVEY (EXISTING SITE CONDITIONS)

PES
Professional Engineering Services, LLC
Civil . Environmental . Structural
P.O. BOX 68573
Renton, WA 98058
Phone: (425) 919-8592

ALI SHASTI, P.E.

	NAME	DATE	
DRAWN BY:	DHB	5/30/05	
CHECKED BY:	MJL	6/30/05	
PARTY CHIEF:	TB	8/29/05	
FIELD BOOK:		1/23/07	

		REVISIONS		NAME	DATE
3					
2	1	ADDED TOPOGRAPHIC & TREE SURVEY TO BOUNDARY	JJB	MJL	3/22/07
DATE	NO.	DESCRIPTION	MADE BY	APPD.	DATE

3/22/07

144

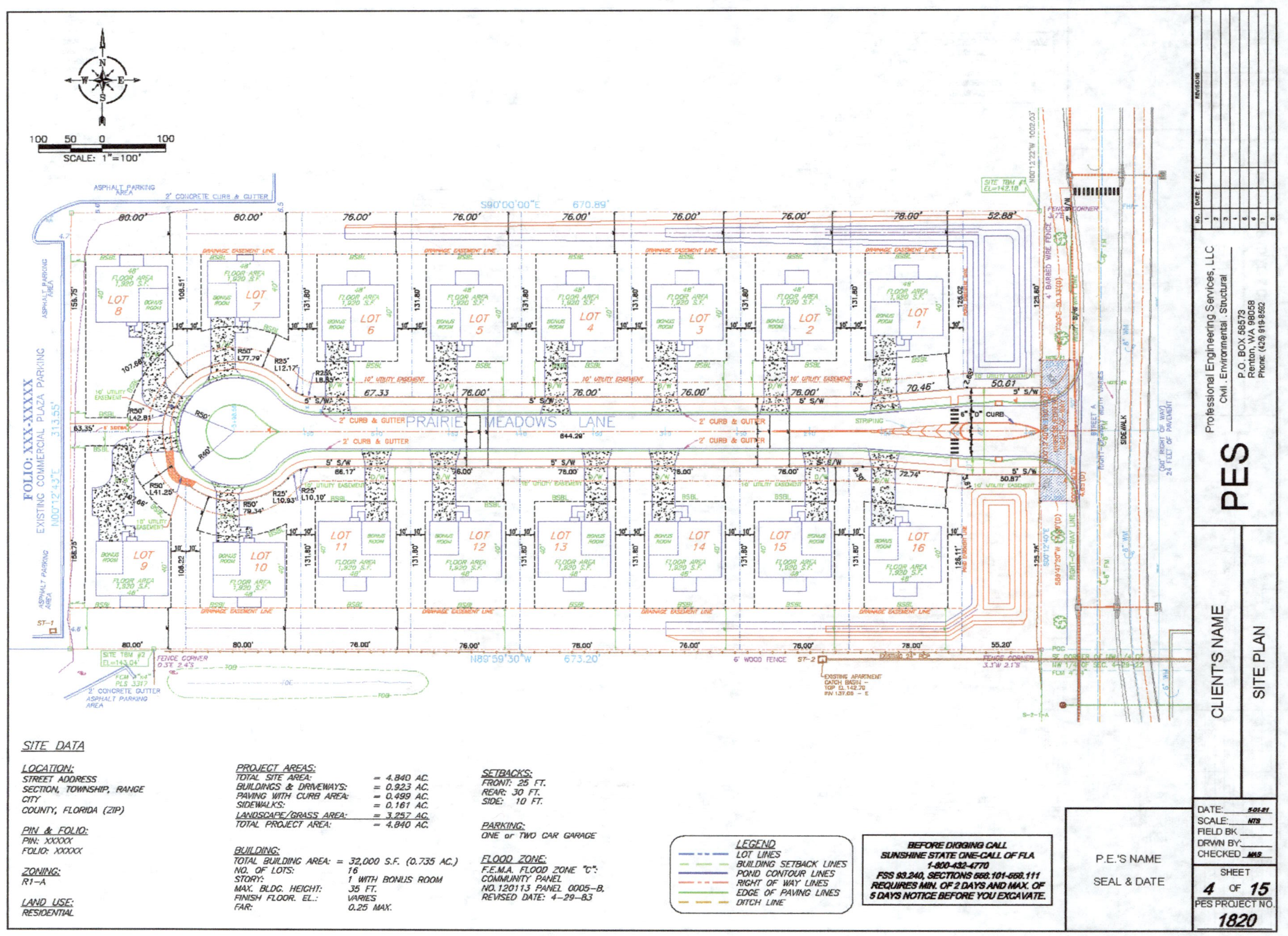

SITE DATA

LOCATION:
STREET ADDRESS
SECTION, TOWNSHIP, RANGE
CITY
COUNTY, FLORIDA (ZIP)

PIN & FOLIO:
PIN: XXXXX
FOLIO: XXXXX

ZONING:
R1-A

LAND USE:
RESIDENTIAL

PROJECT AREAS:
TOTAL SITE AREA: = 4.840 AC.
BUILDINGS & DRIVEWAYS: = 0.923 AC.
PAVING WITH CURB AREA: = 0.499 AC.
SIDEWALKS: = 0.161 AC.
LANDSCAPE/GRASS AREA: = 3.257 AC.
TOTAL PROJECT AREA: = 4.840 AC.

BUILDING:
TOTAL BUILDING AREA: = 32,000 S.F. (0.735 AC.)
NO. OF LOTS: 16
STORY: 1 WITH BONUS ROOM
MAX. BLDG. HEIGHT: 35 FT.
FINISH FLOOR. EL.:   VARIES
FAR: 0.25 MAX.

SETBACKS:
FRONT: 25 FT.
REAR: 30 FT.
SIDE: 10 FT.

PARKING:
ONE or TWO CAR GARAGE

FLOOD ZONE:
F.E.M.A. FLOOD ZONE "C":
COMMUNITY PANEL
NO.120113 PANEL 0005-B,
REVISED DATE: 4-29-83

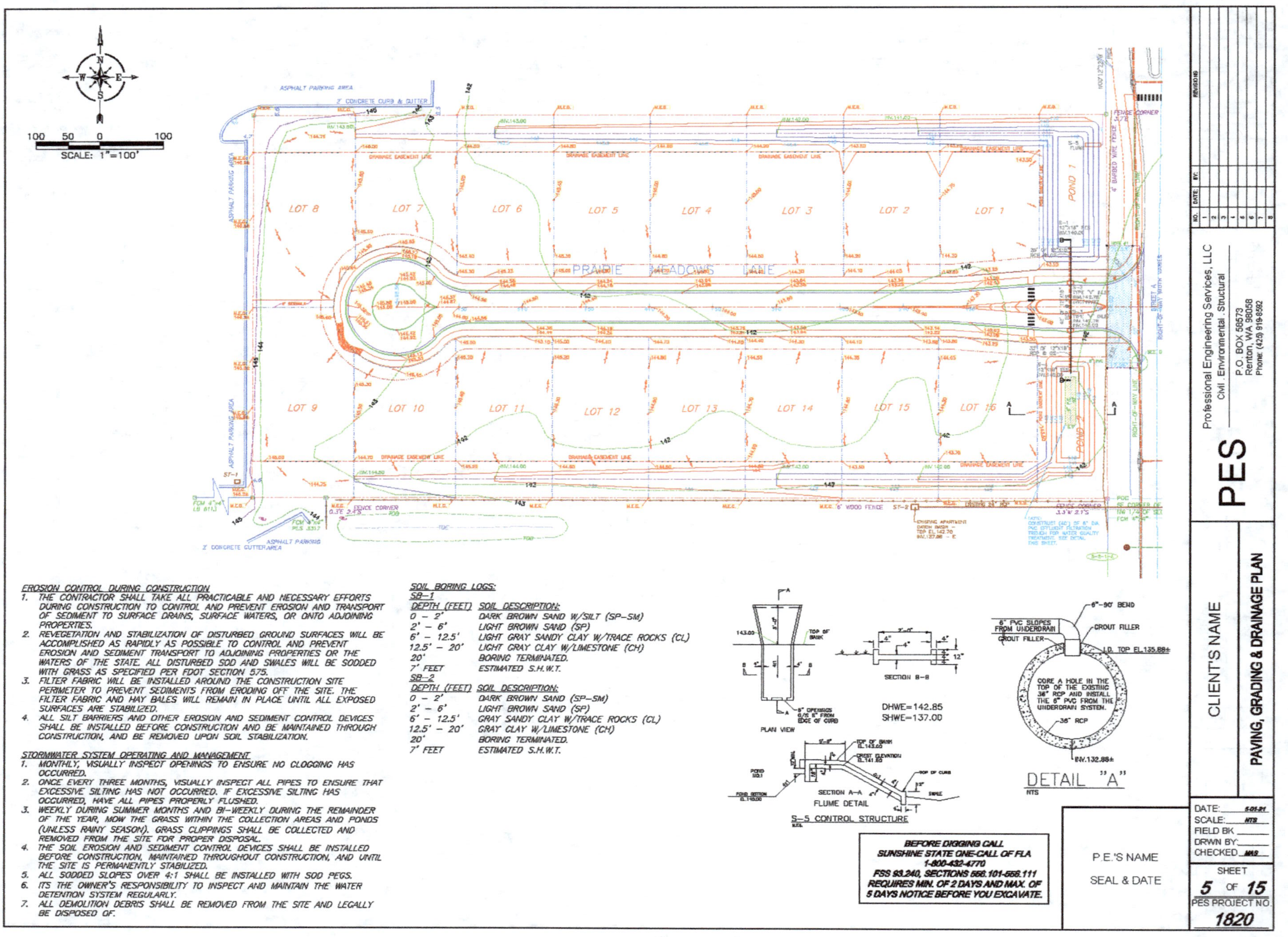

SCALE: 1"=100'
100 50 0 100
Professional Engineering Services, LLC
Civil . Environmental . Structural
P.O. BOX 58673
Renton, WA 98058
Phone (425) 919-8582
PES
CLIENT'S NAME
PAVING, GRADING & DRAINAGE PLAN
DATE:
SCALE:
FIELD BK
DRWN BY:
CHECKED:
SHEET 5 OF 15
PES PROJECT NO. 1820
P.E.'S NAME
SEAL & DATE
DETAIL "A"
NTS
6"-90° BEND
6" PVC SLOPES FROM UNDERDRAIN
GROUT FILLER
I.D. TOP EL.135.88±
CORE A HOLE IN THE TOP OF THE EXISTING 36" RCP AND INSTALL THE 6" PVC FROM THE UNDERDRAIN SYSTEM.
36" RCP
INV.132.88±
SECTION B-B
DHWE=142.85
SHWE=137.00
PLAN VIEW
SECTION A-A
FLUME DETAIL
S-5 CONTROL STRUCTURE
BEFORE DIGGING CALL
SUNSHINE STATE ONE-CALL OF FLA
1-800-432-4770
FSS 93.240, SECTIONS 556.101-556.111
REQUIRES MIN. OF 2 DAYS AND MAX. OF
5 DAYS NOTICE BEFORE YOU EXCAVATE.

LOT 1   LOT 2   LOT 3   LOT 4   LOT 5   LOT 6   LOT 7   LOT 8
LOT 9   LOT 10   LOT 11   LOT 12   LOT 13   LOT 14   LOT 15   LOT 16
PRAIRIE MEADOWS LANE
POND 1
ASPHALT PARKING AREA
2' CONCRETE CURB & GUTTER
2' CONCRETE GUTTER AREA
DRAINAGE EASEMENT LINE

SOIL BORING LOGS:
SB-1
DEPTH (FEET)   SOIL DESCRIPTION:
0 - 2'   DARK BROWN SAND W/SILT (SP-SM)
2' - 6'   LIGHT BROWN SAND (SP)
6' - 12.5' LIGHT GRAY SANDY CLAY W/TRACE ROCKS (CL)
12.5' - 20' LIGHT GRAY CLAY W/LIMESTONE (CH)
20'   BORING TERMINATED.
7' FEET   ESTIMATED S.H.W.T.
SB-2
DEPTH (FEET)   SOIL DESCRIPTION:
0 - 2'   DARK BROWN SAND (SP-SM)
2' - 6'   LIGHT BROWN SAND (SP)
6' - 12.5' GRAY SANDY CLAY W/TRACE ROCKS (CL)
12.5' - 20' GRAY CLAY W/LIMESTONE (CH)
20'   BORING TERMINATED.
7' FEET   ESTIMATED S.H.W.T.

EROSION CONTROL DURING CONSTRUCTION:
1. THE CONTRACTOR SHALL TAKE ALL PRACTICABLE AND NECESSARY EFFORTS
DURING CONSTRUCTION TO CONTROL AND PREVENT EROSION AND TRANSPORT
OF SEDIMENT TO SURFACE DRAINS, SURFACE WATERS, OR ONTO ADJOINING
PROPERTIES.
2. REVEGETATION AND STABILIZATION OF DISTURBED GROUND SURFACES WILL BE
ACCOMPLISHED AS RAPIDLY AS POSSIBLE TO CONTROL AND PREVENT
EROSION AND SEDIMENT TRANSPORT TO ADJOINING PROPERTIES OR THE
WATERS OF THE STATE. ALL DISTURBED SOD AND SWALES WILL BE SODDED
WITH GRASS AS SPECIFIED PER FDOT SECTION 575.
3. FILTER FABRIC WILL BE INSTALLED AROUND THE CONSTRUCTION SITE
PERIMETER TO PREVENT SEDIMENTS FROM ERODING OFF THE SITE. THE
FILTER FABRIC AND HAY BALES WILL REMAIN IN PLACE UNTIL ALL EXPOSED
SURFACES ARE STABILIZED.
4. ALL SILT BARRIERS AND OTHER EROSION AND SEDIMENT CONTROL DEVICES
SHALL BE INSTALLED BEFORE CONSTRUCTION AND BE MAINTAINED THROUGH
CONSTRUCTION, AND BE REMOVED UPON SOIL STABILIZATION.

STORMWATER SYSTEM OPERATING AND MANAGEMENT:
1. MONTHLY, VISUALLY INSPECT OPENINGS TO ENSURE NO CLOGGING HAS
OCCURRED.
2. ONCE EVERY THREE MONTHS, VISUALLY INSPECT ALL PIPES TO ENSURE THAT
EXCESSIVE SILTING HAS NOT OCCURRED. IF EXCESSIVE SILTING HAS
OCCURRED, HAVE ALL PIPES PROPERLY FLUSHED.
3. WEEKLY DURING SUMMER MONTHS AND BI-WEEKLY DURING THE REMAINDER
OF THE YEAR, MOW THE GRASS WITHIN THE COLLECTION AREAS AND PONDS
(UNLESS RAINY SEASON). GRASS CLIPPINGS SHALL BE COLLECTED AND
REMOVED FROM THE SITE FOR PROPER DISPOSAL.
4. THE SOIL EROSION AND SEDIMENT CONTROL DEVICES SHALL BE INSTALLED
BEFORE CONSTRUCTION, MAINTAINED THROUGHOUT CONSTRUCTION, AND UNTIL
THE SITE IS PERMANENTLY STABILIZED.
5. ALL SODDED SLOPES OVER 4:1 SHALL BE INSTALLED WITH SOD PEGS.
6. ITS THE OWNER'S RESPONSIBILITY TO INSPECT AND MAINTAIN THE WATER
DETENTION SYSTEM REGULARLY.
7. ALL DEMOLITION DEBRIS SHALL BE REMOVED FROM THE SITE AND LEGALLY
BE DISPOSED OF.

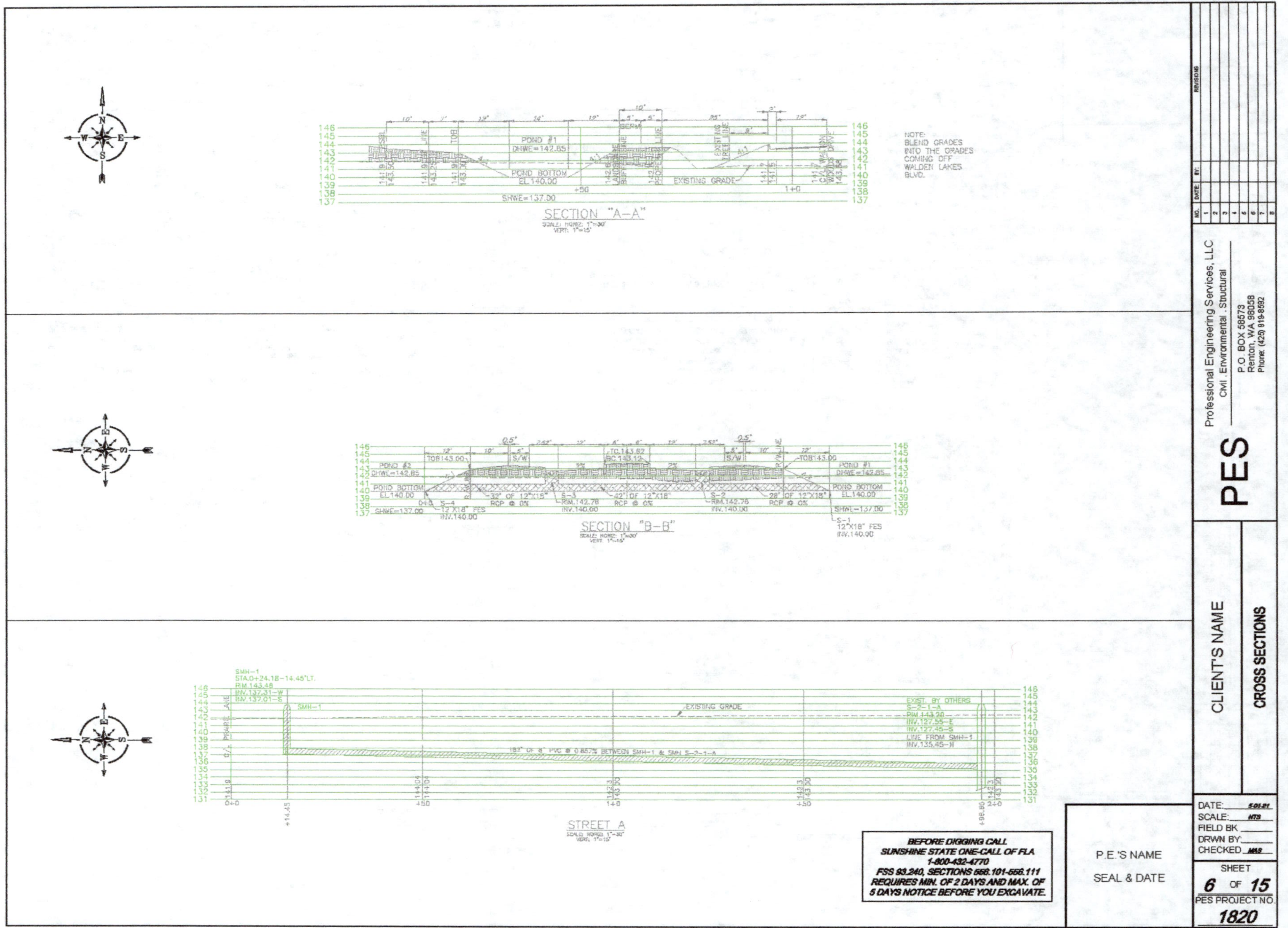

SECTION "A-A"
NOTE:
BLEND GRADES
INTO THE GRADES
COMING OFF
WALDEN LAKES
BLVD.
SECTION "B-B"
STREET A
CLIENT'S NAME
CROSS SECTIONS
PES
Professional Engineering Services, LLC
Civil . Environmental . Structural
P.O. BOX 58573
Renton, WA 98058
Phone: (425) 919-8592
P.E.'S NAME
SEAL & DATE
DATE:
SCALE: NTS
FIELD BK
DRWN BY:
CHECKED: MAS
SHEET
6 OF 15
PES PROJECT NO.
1820
BEFORE DIGGING CALL
SUNSHINE STATE ONE-CALL OF FLA.
1-800-432-4770
FSS 93.240, SECTIONS 556.101-556.111
REQUIRES MIN. OF 2 DAYS AND MAX. OF
5 DAYS NOTICE BEFORE YOU EXCAVATE.
147

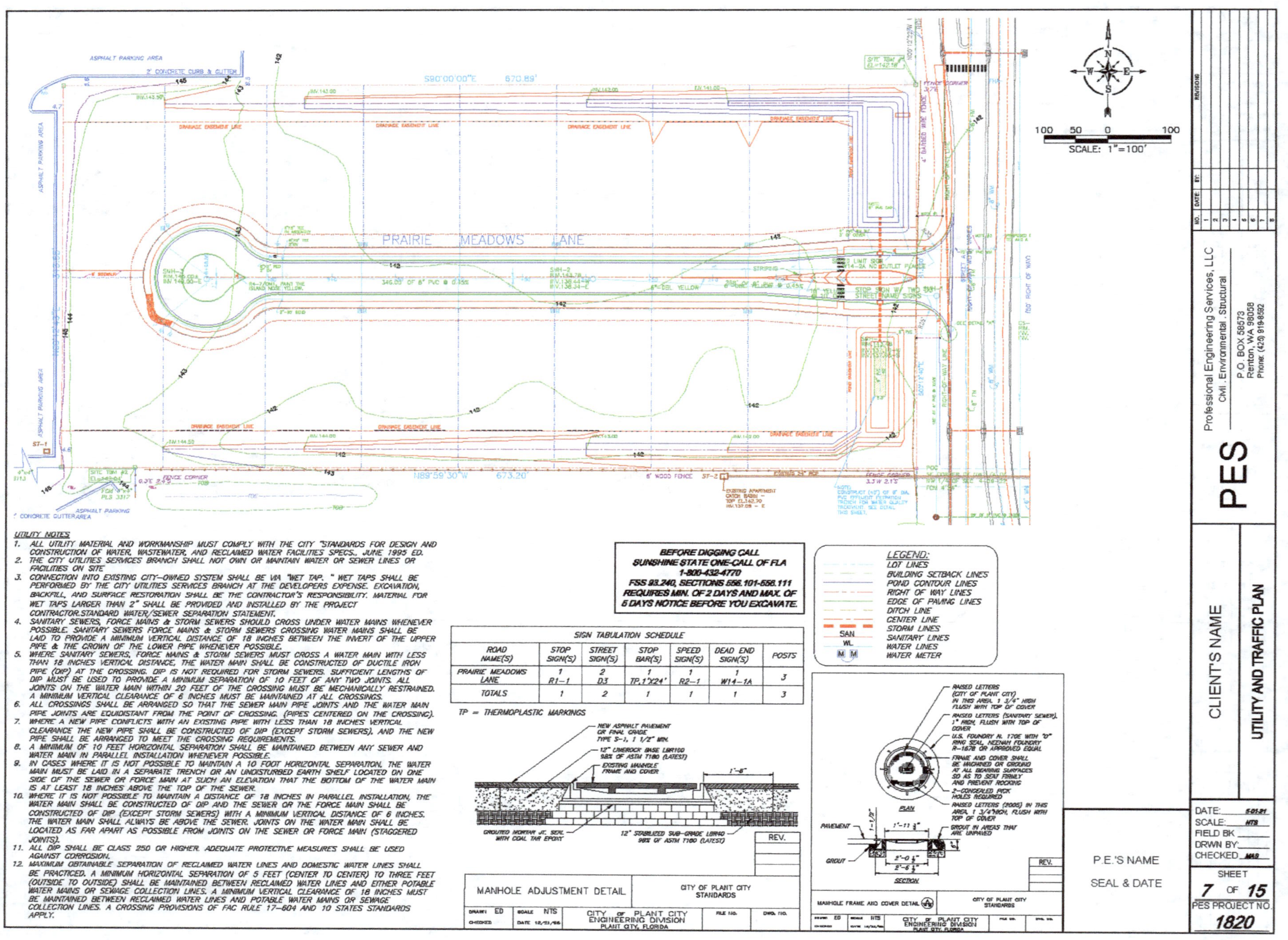

UTILITY NOTES:

1. ALL UTILITY MATERIAL AND WORKMANSHIP MUST COMPLY WITH THE CITY "STANDARDS FOR DESIGN AND CONSTRUCTION OF WATER, WASTEWATER, AND RECLAIMED WATER FACILITIES SPECS., JUNE 1995 ED.
2. THE CITY UTILITIES SERVICES BRANCH SHALL NOT OWN OR MAINTAIN WATER OR SEWER LINES OR FACILITIES ON SITE.
3. CONNECTION INTO EXISTING CITY-OWNED SYSTEM SHALL BE VIA "WET TAP." WET TAPS SHALL BE PERFORMED BY THE CITY UTILITIES SERVICES BRANCH AT THE DEVELOPERS EXPENSE. EXCAVATION, BACKFILL, AND SURFACE RESTORATION SHALL BE THE CONTRACTOR'S RESPONSIBILITY. MATERIAL FOR WET TAPS LARGER THAN 2" SHALL BE PROVIDED AND INSTALLED BY THE PROJECT CONTRACTOR. STANDARD WATER/SEWER SEPARATION STATEMENT.
4. SANITARY SEWERS, FORCE MAINS & STORM SEWERS SHOULD CROSS UNDER WATER MAINS WHENEVER POSSIBLE. SANITARY SEWERS FORCE MAINS & STORM SEWERS CROSSING WATER MAINS SHALL BE LAID TO PROVIDE A MINIMUM VERTICAL DISTANCE OF 18 INCHES BETWEEN THE INVERT OF THE UPPER PIPE & THE CROWN OF THE LOWER PIPE WHENEVER POSSIBLE.
5. WHERE SANITARY SEWERS, FORCE MAINS & STORM SEWERS MUST CROSS A WATER MAIN WITH LESS THAN 18 INCHES VERTICAL DISTANCE, THE WATER MAIN SHALL BE CONSTRUCTED OF DUCTILE IRON PIPE (DIP) AT THE CROSSING. DIP IS NOT REQUIRED FOR STORM SEWERS. SUFFICIENT LENGTHS OF DIP MUST BE USED TO PROVIDE A MINIMUM SEPARATION OF 10 FEET OF ANY TWO JOINTS. ALL JOINTS ON THE WATER MAIN WITHIN 20 FEET OF THE CROSSING MUST BE MECHANICALLY RESTRAINED. A MINIMUM VERTICAL CLEARANCE OF 6 INCHES MUST BE MAINTAINED AT ALL CROSSINGS.
6. ALL CROSSINGS SHALL BE ARRANGED SO THAT THE SEWER MAIN PIPE JOINTS AND THE WATER MAIN PIPE JOINTS ARE EQUIDISTANT FROM THE POINT OF CROSSING. (PIPES CENTERED ON THE CROSSING).
7. WHERE A NEW PIPE CONFLICTS WITH AN EXISTING PIPE WITH LESS THAN 18 INCHES VERTICAL CLEARANCE THE NEW PIPE SHALL BE CONSTRUCTED OF DIP (EXCEPT STORM SEWERS), AND THE NEW PIPE SHALL BE ARRANGED TO MEET THE CROSSING REQUIREMENTS.
8. A MINIMUM OF 10 FEET HORIZONTAL SEPARATION SHALL BE MAINTAINED BETWEEN ANY SEWER AND WATER MAIN IN PARALLEL INSTALLATION WHENEVER POSSIBLE.
9. IN CASES WHERE IT IS NOT POSSIBLE TO MAINTAIN A 10 FOOT HORIZONTAL SEPARATION, THE WATER MAIN MUST BE LAID IN A SEPARATE TRENCH OR AN UNDISTURBED EARTH SHELF LOCATED ON ONE SIDE OF THE SEWER OR FORCE MAIN AT SUCH AN ELEVATION THAT THE BOTTOM OF THE WATER MAIN IS AT LEAST 18 INCHES ABOVE THE TOP OF THE SEWER.
10. WHERE IT IS NOT POSSIBLE TO MAINTAIN A DISTANCE OF 18 INCHES IN PARALLEL INSTALLATION, THE WATER MAIN SHALL BE CONSTRUCTED OF DIP AND THE SEWER OR THE FORCE MAIN SHALL BE CONSTRUCTED OF DIP (EXCEPT STORM SEWERS) WITH A MINIMUM VERTICAL DISTANCE OF 6 INCHES. THE WATER MAIN SHALL ALWAYS BE ABOVE THE SEWER. JOINTS ON THE WATER MAIN SHALL BE LOCATED AS FAR APART AS POSSIBLE FROM JOINTS ON THE SEWER OR FORCE MAIN (STAGGERED JOINTS).
11. ALL DIP SHALL BE CLASS 250 OR HIGHER. ADEQUATE PROTECTIVE MEASURES SHALL BE USED AGAINST CORROSION.
12. MAXIMUM OBTAINABLE SEPARATION OF RECLAIMED WATER LINES AND DOMESTIC WATER LINES SHALL BE PRACTICED. A MINIMUM HORIZONTAL SEPARATION OF 5 FEET (CENTER TO CENTER) TO THREE FEET (OUTSIDE TO OUTSIDE) SHALL BE MAINTAINED BETWEEN RECLAIMED WATER LINES AND EITHER POTABLE WATER MAINS OR SEWAGE COLLECTION LINES. A MINIMUM VERTICAL CLEARANCE OF 18 INCHES MUST BE MAINTAINED BETWEEN RECLAIMED WATER LINES AND POTABLE WATER MAINS OR SEWAGE COLLECTION LINES. A CROSSING PROVISIONS OF FAC RULE 17-604 AND 10 STATES STANDARDS APPLY.

BEFORE DIGGING CALL
SUNSHINE STATE ONE-CALL OF FLA
1-800-432-4770
FSS 93.240, SECTIONS 556.101-556.111
REQUIRES MIN. OF 2 DAYS AND MAX. OF
5 DAYS NOTICE BEFORE YOU EXCAVATE.

SIGN TABULATION SCHEDULE

ROAD NAME(S)	STOP SIGN(S)	STREET SIGN(S)	STOP BAR(S)	SPEED SIGN(S)	DEAD END SIGN(S)	POSTS
PRAIRIE MEADOWS LANE	1 / R1-1	2 / D3	1 / TP,1'X24'	1 / R2-1	1 / W14-1A	3
TOTALS	1	2	1	1	1	3

TP = THERMOPLASTIC MARKINGS

LEGEND:
- LOT LINES
- BUILDING SETBACK LINES
- POND CONTOUR LINES
- RIGHT OF WAY LINES
- EDGE OF PAVING LINES
- DITCH LINE
- CENTER LINE
- STORM LINES
- SAN SANITARY LINES
- WL WATER LINES
- M M WATER METER

MANHOLE ADJUSTMENT DETAIL

NEW ASPHALT PAVEMENT OR FINAL GRADE TYPE S-1, 1 1/2" MIN.
12" LIMEROCK BASE LBR100 98% OF ASTM T180 (LATEST)
EXISTING MANHOLE FRAME AND COVER
GROUTED MORTAR JT. SEAL WITH COAL TAR EPOXY
12" STABILIZED SUB-GRADE LBR40 98% OF ASTM T180 (LATEST)

CITY OF PLANT CITY STANDARDS

DRAWN ED   SCALE NTS
CHECKED   DATE 12/21/96
CITY OF PLANT CITY ENGINEERING DIVISION
PLANT CITY, FLORIDA

MANHOLE FRAME AND COVER DETAIL

- RAISED LETTERS (CITY OF PLANT CITY) IN THIS AREA, 1 3/4" HIGH FLUSH WITH TOP OF COVER
- RAISED LETTERS (SANITARY SEWER), 1" HIGH, FLUSH WITH TOP OF COVER
- U.S. FOUNDRY N. 170E WITH "O" RING SEAL, NEENAH FOUNDRY R-1678 OR APPROVED EQUAL
- FRAME AND COVER SHALL BE MACHINED OR GROUND AT ALL BEARING SURFACES SO AS TO SEAT FIRMLY AND PREVENT ROCKING
- 2-CONCEALED PICK HOLES REQUIRED
- RAISED LETTERS (2005) IN THIS AREA, 1 3/4"HIGH, FLUSH WITH TOP OF COVER
- GROUT IN AREAS THAT ARE UNPAVED
PLAN
SECTION
PAVEMENT
GROUT

CITY OF PLANT CITY STANDARDS
CITY OF PLANT CITY ENGINEERING DIVISION
PLANT CITY, FLORIDA

Title Block

PES
Professional Engineering Services, LLC
Civil, Environmental, Structural
P.O. BOX 58573
Renton, WA 98058
Phone: (425) 8198592

CLIENT'S NAME

UTILITY AND TRAFFIC PLAN

DATE: 5-01-21
SCALE: NTS
FIELD BK
DRWN BY:
CHECKED MAS

P.E.'S NAME
SEAL & DATE

SHEET 7 OF 15
PES PROJECT NO. 1820

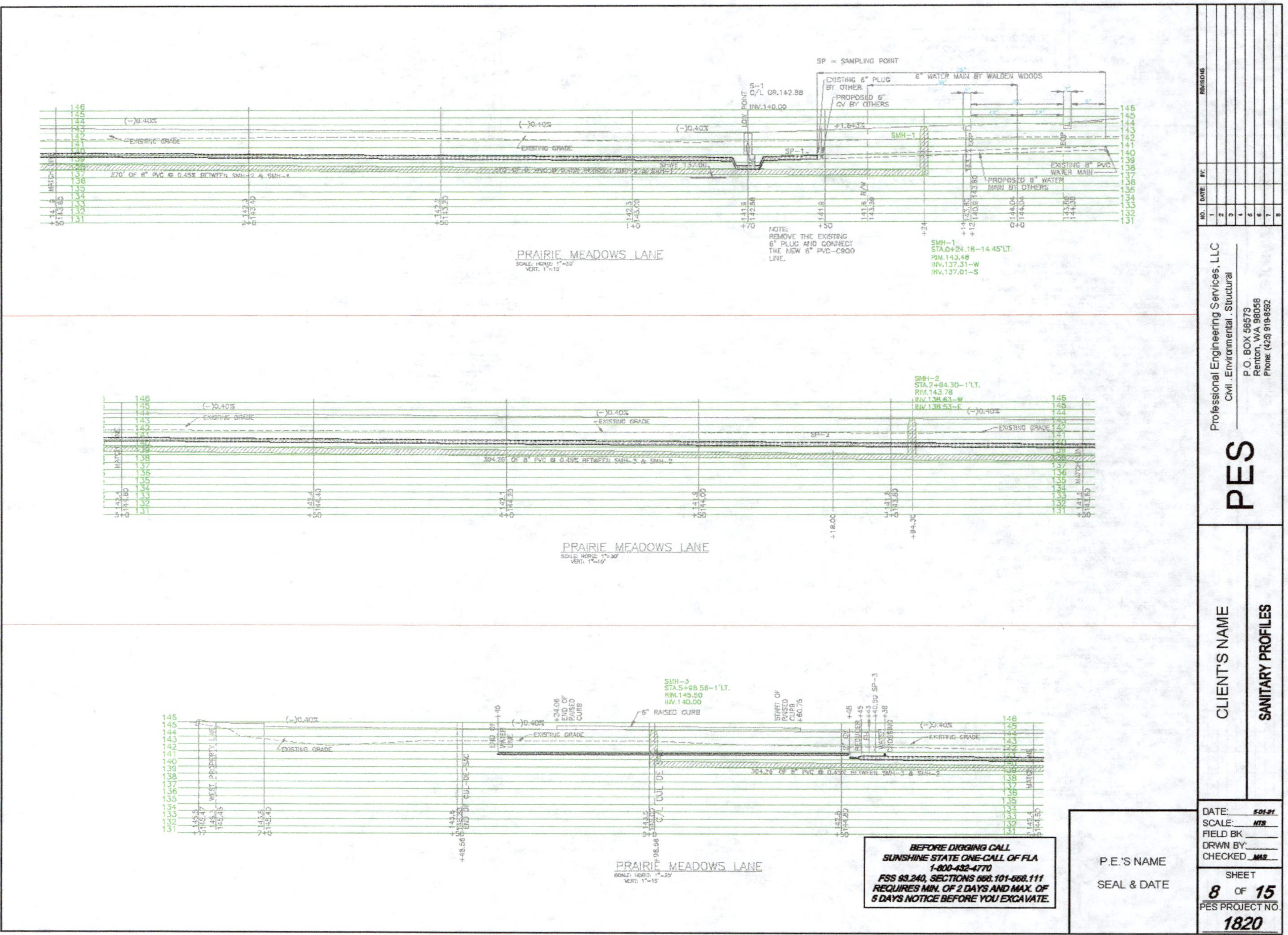

SITE DESIGN & ENGINEERING

PES
Professional Engineering Services, LLC
Civil . Environmental . Structural
P. O. BOX 58573
Renton, WA 98058
Phone. (425) 919-8592

CLIENT'S NAME
SANITARY PROFILES

REVISIONS
NO. DATE BY.

P.E.'S NAME
SEAL & DATE

DATE: 5-01-21
SCALE: NTS
FIELD BK
DRWN BY:
CHECKED MAS
SHEET
8 OF 15
PES PROJECT NO.
1820

SP = SAMPLING POINT

PRAIRIE MEADOWS LANE
SCALE: HORIZ 1"=30'
VERT: 1"=10'

EXISTING 6" PLUG BY OTHER
PROPOSED 6" GV BY OTHERS
8" WATER MAIN BY WALDEN WOODS
EXISTING GRADE
S-1 C/L GR.142.88
JOIN POINT INV.140.00
270' OF 8" PVC @ 0.45% BETWEEN SMH-2 & SMH-1
SMH 137.00
SP-1
SMH-1
PROPOSED 8" WATER MAIN BY OTHERS
EXISTING 8" PVC WATER MAIN
NOTE: REMOVE THE EXISTING 6" PLUG AND CONNECT THE NEW 8" PVC-C900 LINE.
SMH-1
STA.0+24.18-14.45'LT.
RIM.143.48
INV.137.31-W
INV.137.01-S

PRAIRIE MEADOWS LANE
SCALE: HORIZ 1"=30'
VERT: 1"=10'

SMH-2
STA.2+84.30-1'LT.
RIM.143.78
INV.138.63-W
INV.138.53-E
EXISTING GRADE
304.36' OF 8" PVC @ 0.45% BETWEEN SMH-3 & SMH-2
SP-1

SMH-3
STA.5+98.56-1'LT.
RIM.145.50
INV.140.00
6" RAISED CURB
+34.06 END OF RAISED CURB
START OF RAISED CURB +60.75
SP-3
WEST PROPERTY LINE
WATER LINE
END OF RAISED CURB
C/L CUL-DE-SAC
BAD OF CUL-DE-SAC
EXISTING GRADE
304.26' OF 8" PVC @ 0.45% BETWEEN SMH-3 & SMH-2

PRAIRIE MEADOWS LANE
SCALE: HORIZ 1"=30'
VERT: 1"=15'

BEFORE DIGGING CALL
SUNSHINE STATE ONE-CALL OF FLA
1-800-432-4770
FSS 93.240, SECTIONS 556.101-556.111
REQUIRES MIN. OF 2 DAYS AND MAX. OF
5 DAYS NOTICE BEFORE YOU EXCAVATE.

149

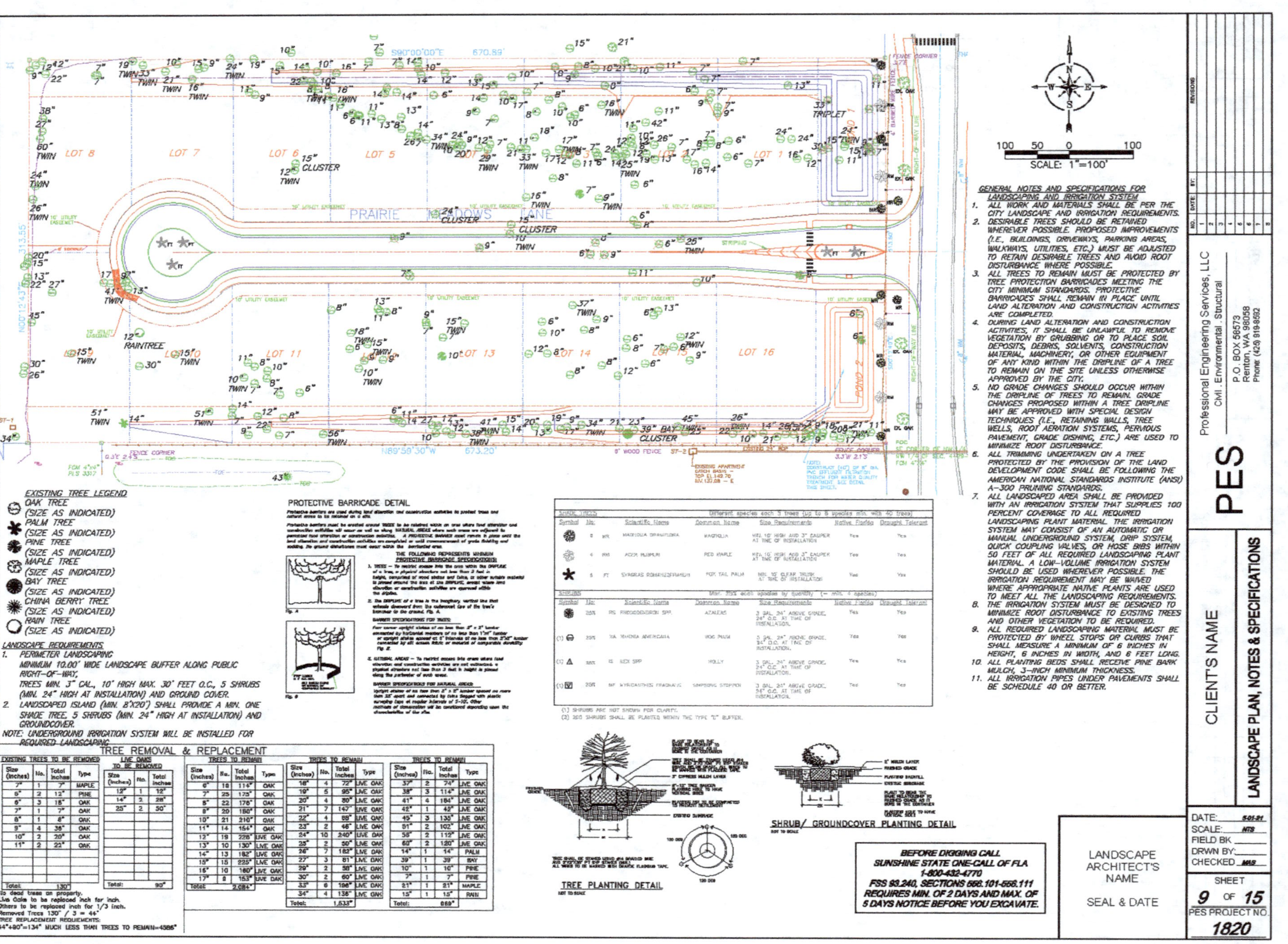

TREE REMOVAL & REPLACEMENT

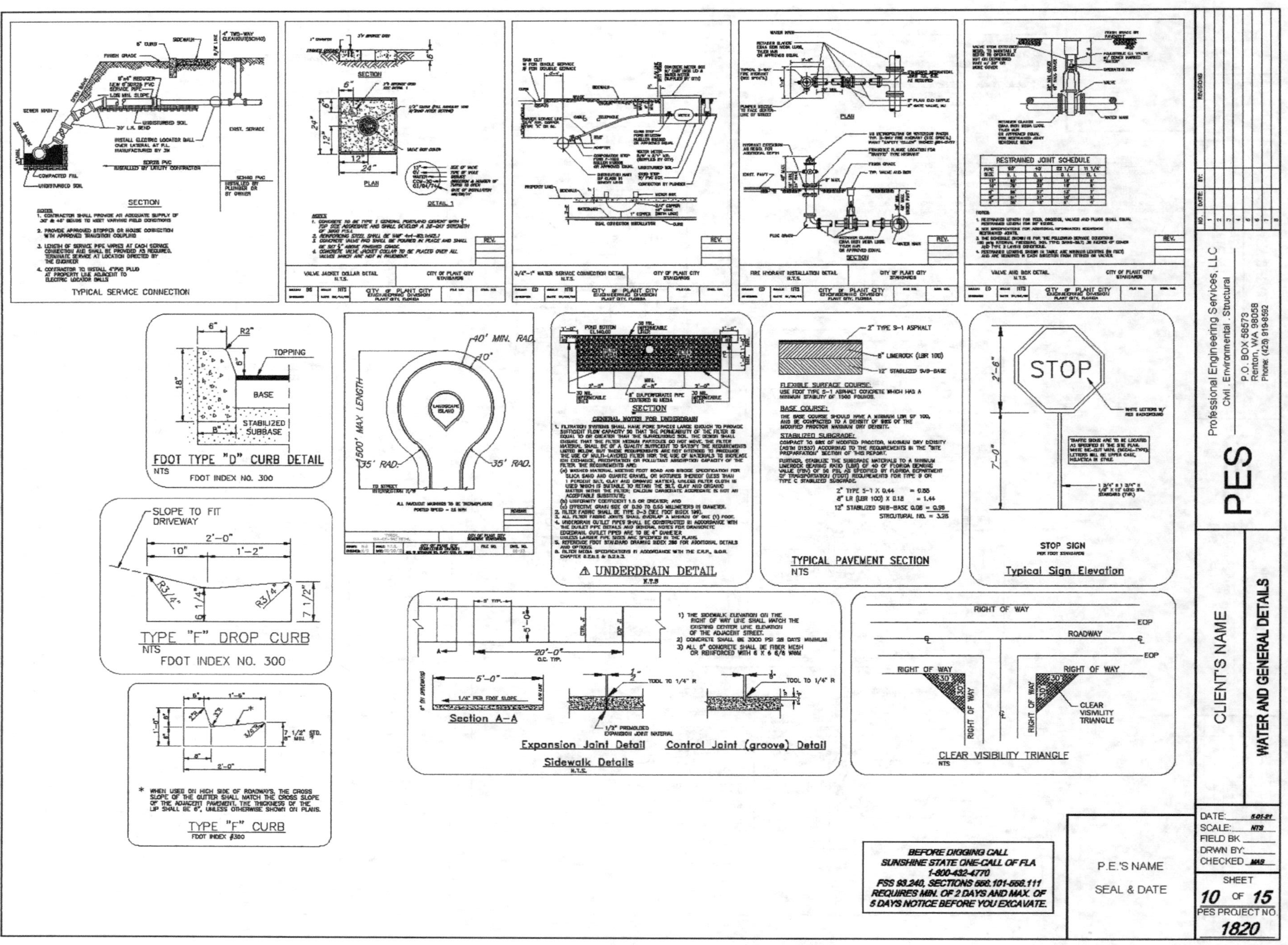

TYPICAL SERVICE CONNECTION
SECTION
DETAIL 1
PLAN

VALVE JACKET COLLAR DETAIL
CITY OF PLANT CITY
STANDARDS
3/4"-1" WATER SERVICE CONNECTION DETAIL
FIRE HYDRANT INSTALLATION DETAIL
VALVE AND BOX DETAIL
RESTRAINED JOINT SCHEDULE
CITY OF PLANT CITY ENGINEERING DIVISION
PLANT CITY, FLORIDA

FDOT TYPE "D" CURB DETAIL
NTS
FDOT INDEX NO. 300

TYPE "F" DROP CURB
NTS
FDOT INDEX NO. 300
SLOPE TO FIT DRIVEWAY

TYPE "F" CURB
FDOT INDEX #300

500' MAX LENGTH
40' MIN. RAD.
35' RAD.
35' RAD.
LANDSCAPE ISLAND
UNDERDRAIN DETAIL
N.T.S
SECTION
GENERAL NOTES FOR UNDERDRAIN

TYPICAL PAVEMENT SECTION
NTS
2" TYPE S-1 ASPHALT
8" LIMEROCK (LBR 100)
12" STABILIZED SUB-BASE
FLEXIBLE SURFACE COURSE:
BASE COURSE:
STABILIZED SUBGRADE:

STOP
STOP SIGN
PER FDOT STANDARDS
Typical Sign Elevation

Section A-A
Expansion Joint Detail   Control Joint (groove) Detail
Sidewalk Details
N.T.S.

RIGHT OF WAY
ROADWAY
RIGHT OF WAY
CLEAR VISIBILITY TRIANGLE
CLEAR VISIBILITY TRIANGLE
NTS
EOP

PES
Professional Engineering Services, LLC
Civil . Environmental . Structural
P.O. BOX 58573
Renton, WA 98058
Phone (425) 919-8592

CLIENT'S NAME
WATER AND GENERAL DETAILS

DATE: 5-01-01
SCALE: NTS
FIELD BK
DRWN BY:
CHECKED MAB
SHEET 10 OF 15
PES PROJECT NO. 1820

P.E.'S NAME
SEAL & DATE

BEFORE DIGGING CALL
SUNSHINE STATE ONE-CALL OF FLA
1-800-432-4770
FSS 93.240, SECTIONS 556.101-556.111
REQUIRES MIN. OF 2 DAYS AND MAX. OF
5 DAYS NOTICE BEFORE YOU EXCAVATE.

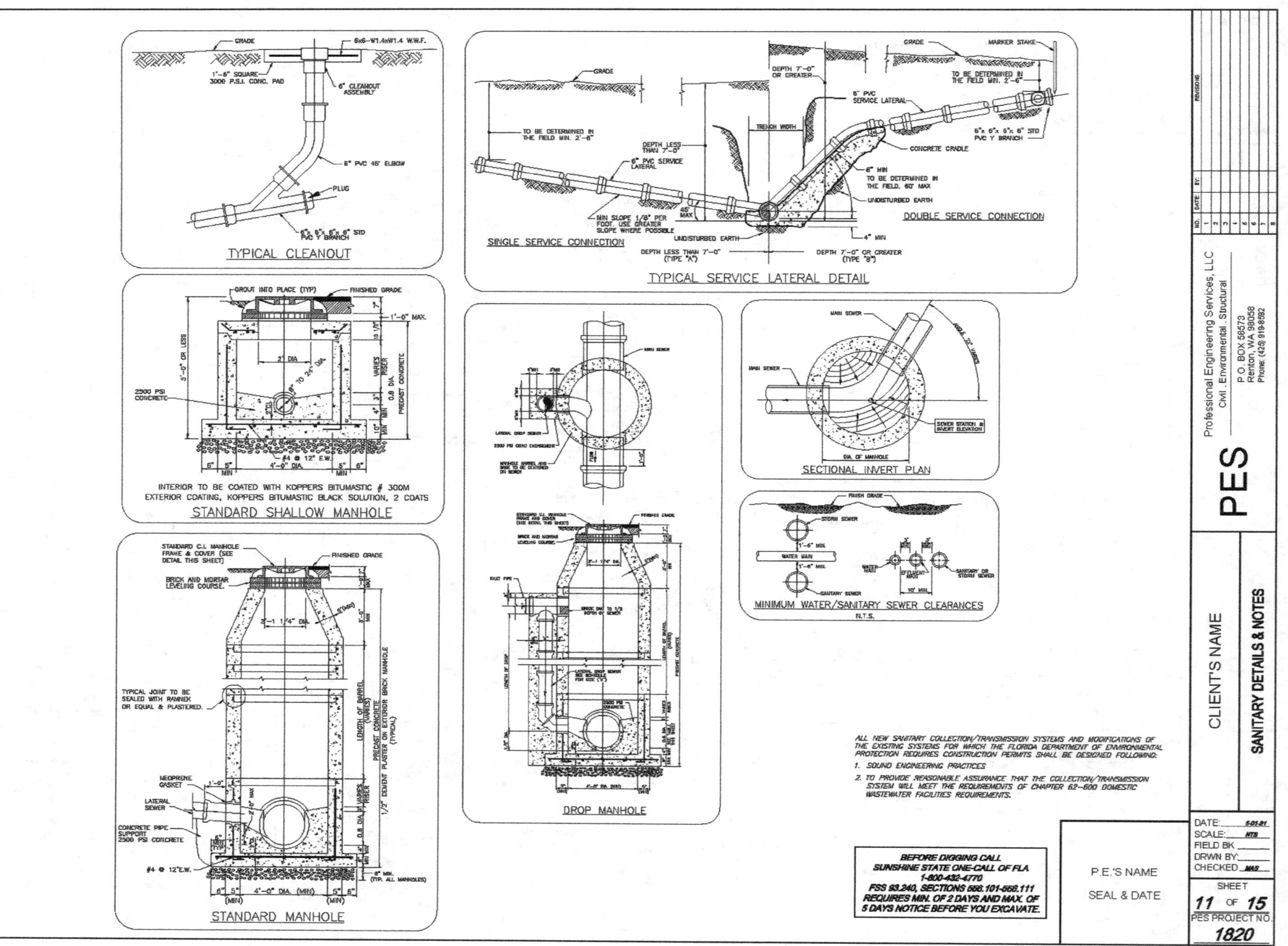
TYPICAL CLEANOUT
GRADE
6x6-W1.4xW1.4 W.W.F.
1'-6" SQUARE
3000 P.S.I. CONC. PAD
6" CLEANOUT ASSEMBLY
6" PVC 45° ELBOW
PLUG
6"x 6"x 6" STD PVC Y BRANCH

TYPICAL SERVICE LATERAL DETAIL
SINGLE SERVICE CONNECTION
DOUBLE SERVICE CONNECTION
GRADE
MARKER STAKE
DEPTH 7'-0" OR GREATER
TO BE DETERMINED IN THE FIELD MIN. 2'-6"
6" PVC SERVICE LATERAL
6"x 6"x 6"x 6" STD PVC Y BRANCH
CONCRETE CRADLE
TO BE DETERMINED IN THE FIELD MIN. 2'-6"
DEPTH LESS THAN 7'-0"
6" PVC SERVICE LATERAL
MIN SLOPE 1/8" PER FOOT. USE GREATER SLOPE WHERE POSSIBLE
45° MAX
UNDISTURBED EARTH
6" MIN TO BE DETERMINED IN THE FIELD, 60" MAX
UNDISTURBED EARTH
4" MIN
TRENCH WIDTH
DEPTH LESS THAN 7'-0" (TYPE "A")
DEPTH 7'-0" OR GREATER (TYPE "B")

STANDARD SHALLOW MANHOLE
INTERIOR TO BE COATED WITH KOPPERS BITUMASTIC # 300M
EXTERIOR COATING, KOPPERS BITUMASTIC BLACK SOLUTION, 2 COATS
GROUT INTO PLACE (TYP)
FINISHED GRADE
1'-0" MAX.
5'-0" OR LESS
2' DIA
VARIES RISER
0.8 DIA.
PRECAST CONCRETE
2500 PSI CONCRETE
#4 @ 12" E.W.
4'-0" DIA.

SECTIONAL INVERT PLAN
MAIN SEWER
SEWER STATION & INVERT ELEVATION
DIA. OF MANHOLE
ANGLE "X" VARIES

MINIMUM WATER/SANITARY SEWER CLEARANCES
N.T.S.
FINISH GRADE
STORM SEWER
WATER MAIN
SANITARY SEWER
SANITARY OR STORM SEWER
WATER MAIN
EFFLUENT MAIN
10' MIN
1'-6" MIN

DROP MANHOLE
MAIN SEWER
LATERAL DROP SEWER
2500 PSI CONC ENCASEMENT
STANDARD C.I. MANHOLE FRAME AND COVER (SEE DETAIL THIS SHEET)
FINISHED GRADE
BRICK AND MORTAR LEVELING COURSE
INLET PIPE
LENGTH OF DROP
BRICK ONE TO 1/2 DEPTH OF SEWER
PRECAST CONCRETE
LENGTH OF BARREL
LATERAL INLET SEWER

STANDARD MANHOLE
STANDARD C.I. MANHOLE FRAME & COVER (SEE DETAIL THIS SHEET)
FINISHED GRADE
BRICK AND MORTAR LEVELING COURSE
TYPICAL JOINT TO BE SEALED WITH RAMNEK OR EQUAL & PLASTERED.
NEOPRENE GASKET
LATERAL SEWER
CONCRETE PIPE SUPPORT 2500 PSI CONCRETE
#4 @ 12"E.W.
LENGTH OF BARREL (VARIES)
PRECAST CONCRETE
1/2" CEMENT PLASTER ON EXTERIOR BRICK MANHOLE (TYPICAL)
0.8 DIA. RISER (VARIES)
4'-0" DIA. (MIN)

ALL NEW SANITARY COLLECTION/TRANSMISSION SYSTEMS AND MODIFICATIONS OF
THE EXISTING SYSTEMS FOR WHICH THE FLORIDA DEPARTMENT OF ENVIRONMENTAL
PROTECTION REQUIRES CONSTRUCTION PERMITS SHALL BE DESIGNED FOLLOWING:
1. SOUND ENGINEERING PRACTICES
2. TO PROVIDE REASONABLE ASSURANCE THAT THE COLLECTION/TRANSMISSION
SYSTEM WILL MEET THE REQUIREMENTS OF CHAPTER 62-600 DOMESTIC
WASTEWATER FACILITIES REQUIREMENTS.

BEFORE DIGGING CALL
SUNSHINE STATE ONE-CALL OF FLA.
1-800-432-4770
FSS 93.240, SECTIONS 556.101-556.111
REQUIRES MIN. OF 2 DAYS AND MAX. OF
5 DAYS NOTICE BEFORE YOU EXCAVATE.

Professional Engineering Services, LLC
Civil . Environmental . Structural
P.O. BOX 58573
Renton, WA 98058
Phone: (425) 919-8592
PES
CLIENT'S NAME
SANITARY DETAILS & NOTES
P.E.'S NAME
SEAL & DATE
DATE: 5-05-21
SCALE: NTS
FIELD BK.
DRWN BY:
CHECKED: MAS
SHEET 11 OF 15
PES PROJECT NO. 1820
REVISIONS
NO. DATE: BY:

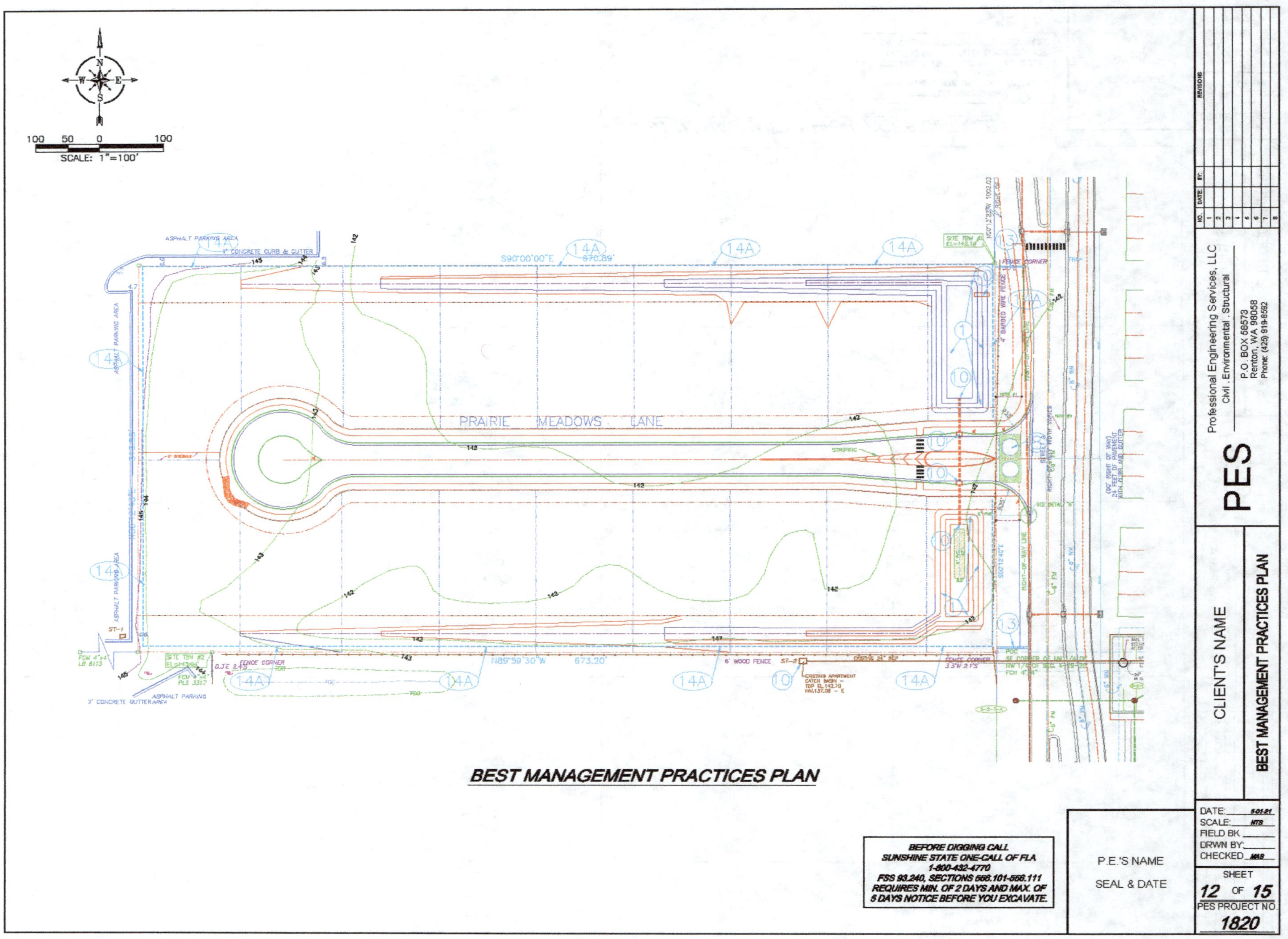

SCALE: 1"=100'
100 50 0 100
N E S W
ASPHALT PARKING AREA
3" CONCRETE CURB & GUTTER
S90°00'00"E 670.89'
14A
14A
14A
PRAIRIE MEADOWS LANE
STRIPING
ASPHALT PARKING AREA
ASPHALT PARKING AREA
2" CONCRETE GUTTER AREA
ASPHALT PARKING
N89°59'30"W 673.20'
6' WOOD FENCE
14A
14A
14A
4' BARBED WIRE FENCE
FENCE CORNER
SITE TBM #1 EL.=142.18
EXISTING APARTMENT
CATCH BASIN
TOP EL.143.70
INV.137.08 - E
EXISTING 24" RCP
FENCE CORNER
SITE TBM #2
FCM 4"x4" PLS 3317
FCM 4"x4" LB 6113
ST-1
BEST MANAGEMENT PRACTICES PLAN
Professional Engineering Services, LLC
Civil . Environmental . Structural
P.O. BOX 68573
Renton, WA 98058
Phone: (425) 919-8592
PES
CLIENT'S NAME
BEST MANAGEMENT PRACTICES PLAN
P.E.'S NAME
SEAL & DATE
DATE: 5-01-21
SCALE: NTS
FIELD BK
DRWN BY:
CHECKED: MAS
SHEET 12 OF 15
PES PROJECT NO. 1820
REVISIONS
NO. DATE: BY:
BEFORE DIGGING CALL
SUNSHINE STATE ONE-CALL OF FLA
1-800-432-4770
FSS 93.240, SECTIONS 668.101-668.111
REQUIRES MIN. OF 2 DAYS AND MAX. OF
5 DAYS NOTICE BEFORE YOU EXCAVATE.

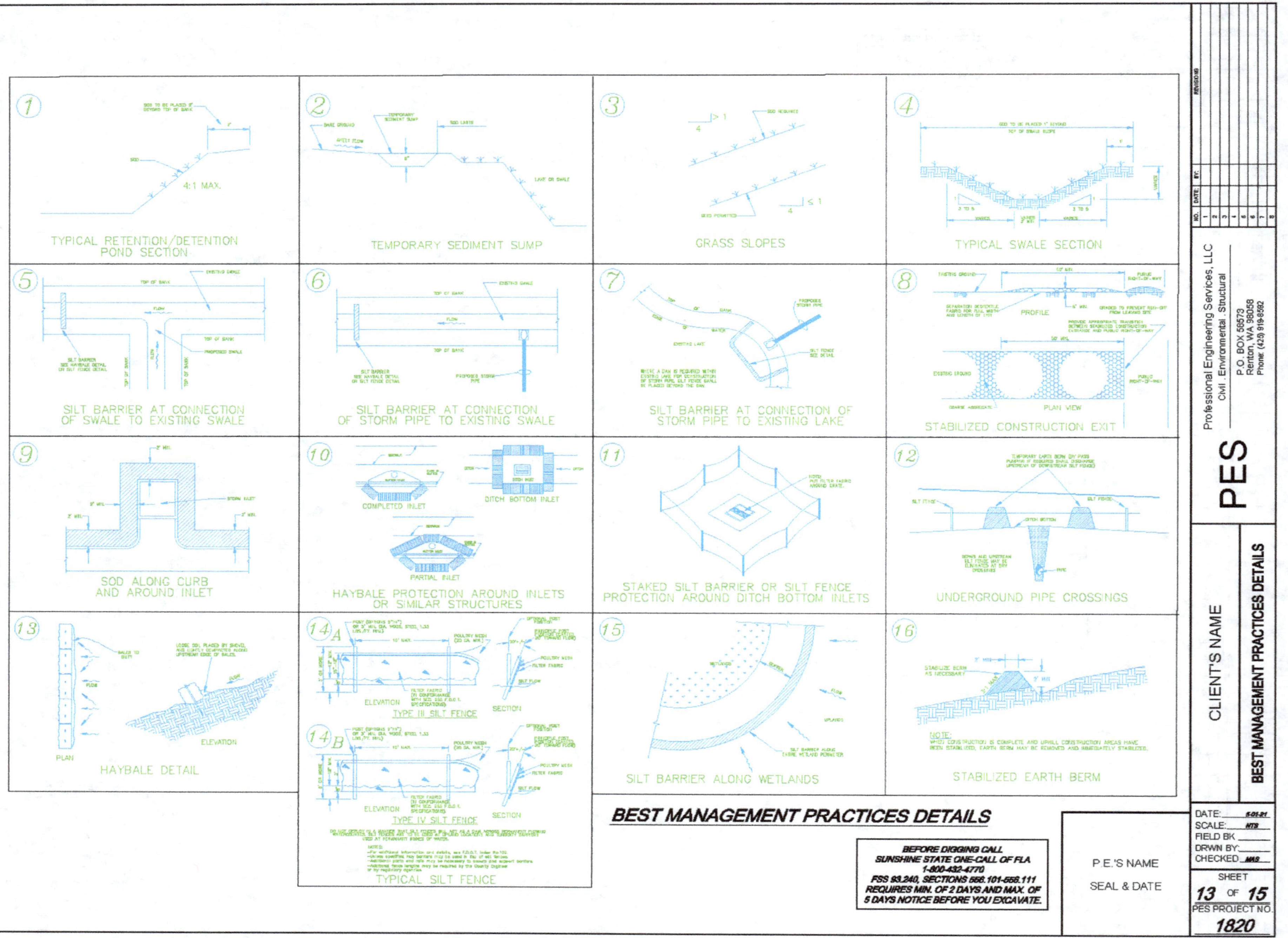
1 TYPICAL RETENTION/DETENTION POND SECTION
4:1 MAX.
2 TEMPORARY SEDIMENT SUMP
3 GRASS SLOPES
4 TYPICAL SWALE SECTION
5 SILT BARRIER AT CONNECTION OF SWALE TO EXISTING SWALE
6 SILT BARRIER AT CONNECTION OF STORM PIPE TO EXISTING SWALE
7 SILT BARRIER AT CONNECTION OF STORM PIPE TO EXISTING LAKE
8 STABILIZED CONSTRUCTION EXIT
PROFILE
PLAN VIEW
9 SOD ALONG CURB AND AROUND INLET
10 COMPLETED INLET
DITCH BOTTOM INLET
PARTIAL INLET
HAYBALE PROTECTION AROUND INLETS OR SIMILAR STRUCTURES
11 STAKED SILT BARRIER OR SILT FENCE PROTECTION AROUND DITCH BOTTOM INLETS
12 UNDERGROUND PIPE CROSSINGS
13 PLAN
ELEVATION
HAYBALE DETAIL
14A ELEVATION
SECTION
TYPE III SILT FENCE
14B ELEVATION
SECTION
TYPE IV SILT FENCE
TYPICAL SILT FENCE
15 SILT BARRIER ALONG WETLANDS
16 STABILIZED EARTH BERM

BEST MANAGEMENT PRACTICES DETAILS

BEFORE DIGGING CALL
SUNSHINE STATE ONE-CALL OF FLA
1-800-432-4770
FSS 93.240, SECTIONS 556.101-556.111
REQUIRES MIN. OF 2 DAYS AND MAX. OF
5 DAYS NOTICE BEFORE YOU EXCAVATE.

Professional Engineering Services, LLC
Civil . Environmental . Structural
P.O. BOX 58673
Renton, WA 98058
Phone (425) 919-8592
PES
CLIENT'S NAME
BEST MANAGEMENT PRACTICES DETAILS
P.E.'S NAME
SEAL & DATE
DATE: 5-01-01
SCALE: NTS
FIELD BK.
DRWN BY:
CHECKED: MAS
SHEET 13 OF 15
PES PROJECT NO. 1820

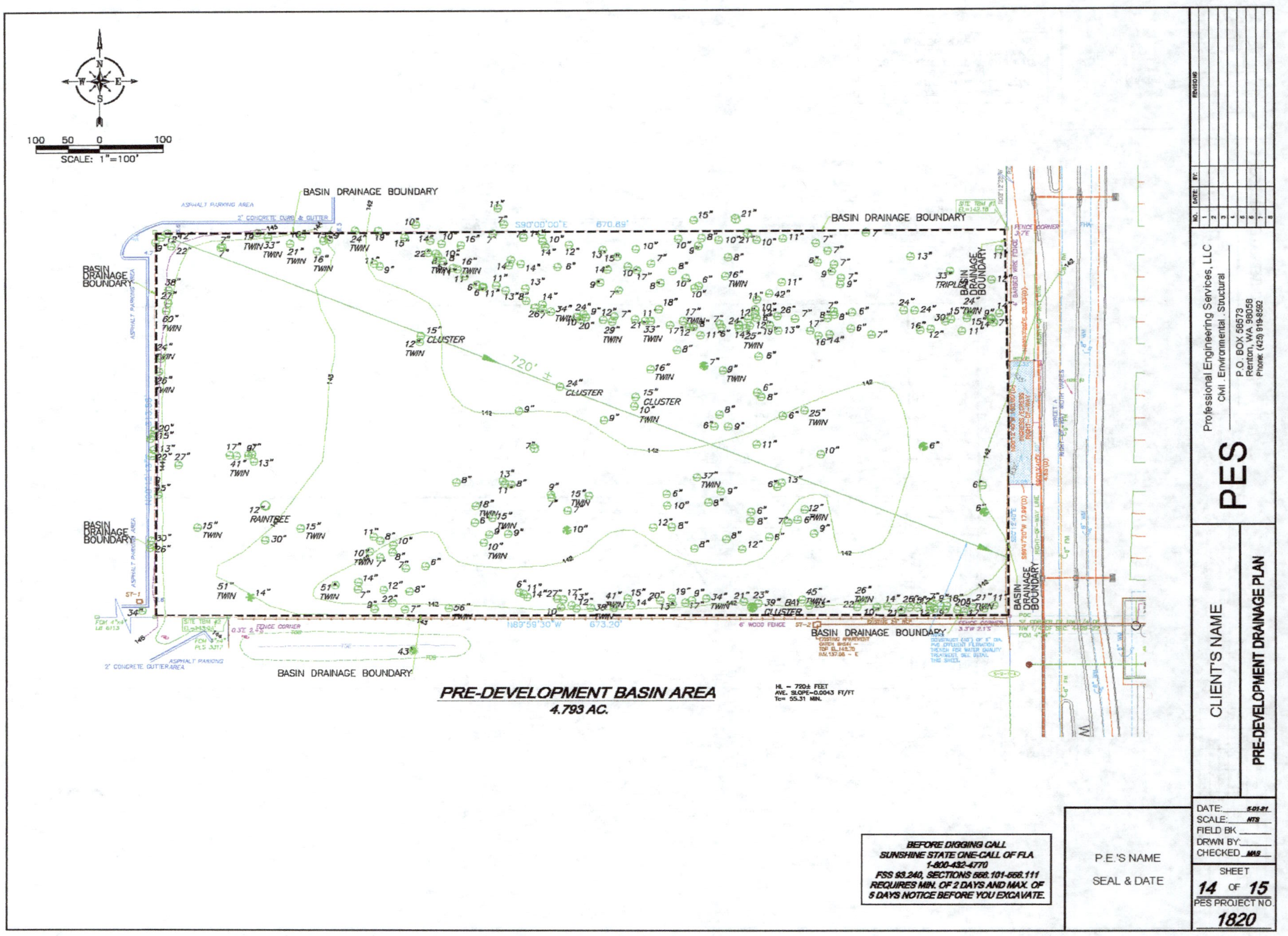
SCALE: 1"=100'
100 50 0 100
BASIN DRAINAGE BOUNDARY
BASIN DRAINAGE BOUNDARY
S90°00'00"E 870.89'
ASPHALT PARKING AREA
2' CONCRETE CURB & GUTTER
BASIN DRAINAGE BOUNDARY
720' ±
CLUSTER
TWIN
TRIPLE
RAINTREE
BASIN DRAINAGE BOUNDARY
ASPHALT PARKING AREA
2' CONCRETE GUTTER AREA
N89°59'30"W 873.20'
6' WOOD FENCE
BASIN DRAINAGE BOUNDARY
BASIN DRAINAGE BOUNDARY
PRE-DEVELOPMENT BASIN AREA
4.793 AC.
HL = 720± FEET
AVE. SLOPE=0.0043 FT/FT
Tc= 55.31 MIN.
CLIENT'S NAME
PRE-DEVELOPMENT DRAINAGE PLAN
PES
Professional Engineering Services, LLC
Civil . Environmental . Structural
P.O. BOX 58573
Renton, WA 98058
Phone: (425) 919-8592
REVISIONS
NO. DATE BY.
DATE: 5-01-21
SCALE: NTS
FIELD BK
DRWN BY:
CHECKED MAS
SHEET
14 OF 15
PES PROJECT NO.
1820
P.E.'S NAME
SEAL & DATE
BEFORE DIGGING CALL
SUNSHINE STATE ONE-CALL OF FLA
1-800-432-4770
FSS §3.240, SECTIONS 556.101-556.111
REQUIRES MIN. OF 2 DAYS AND MAX. OF
5 DAYS NOTICE BEFORE YOU EXCAVATE.

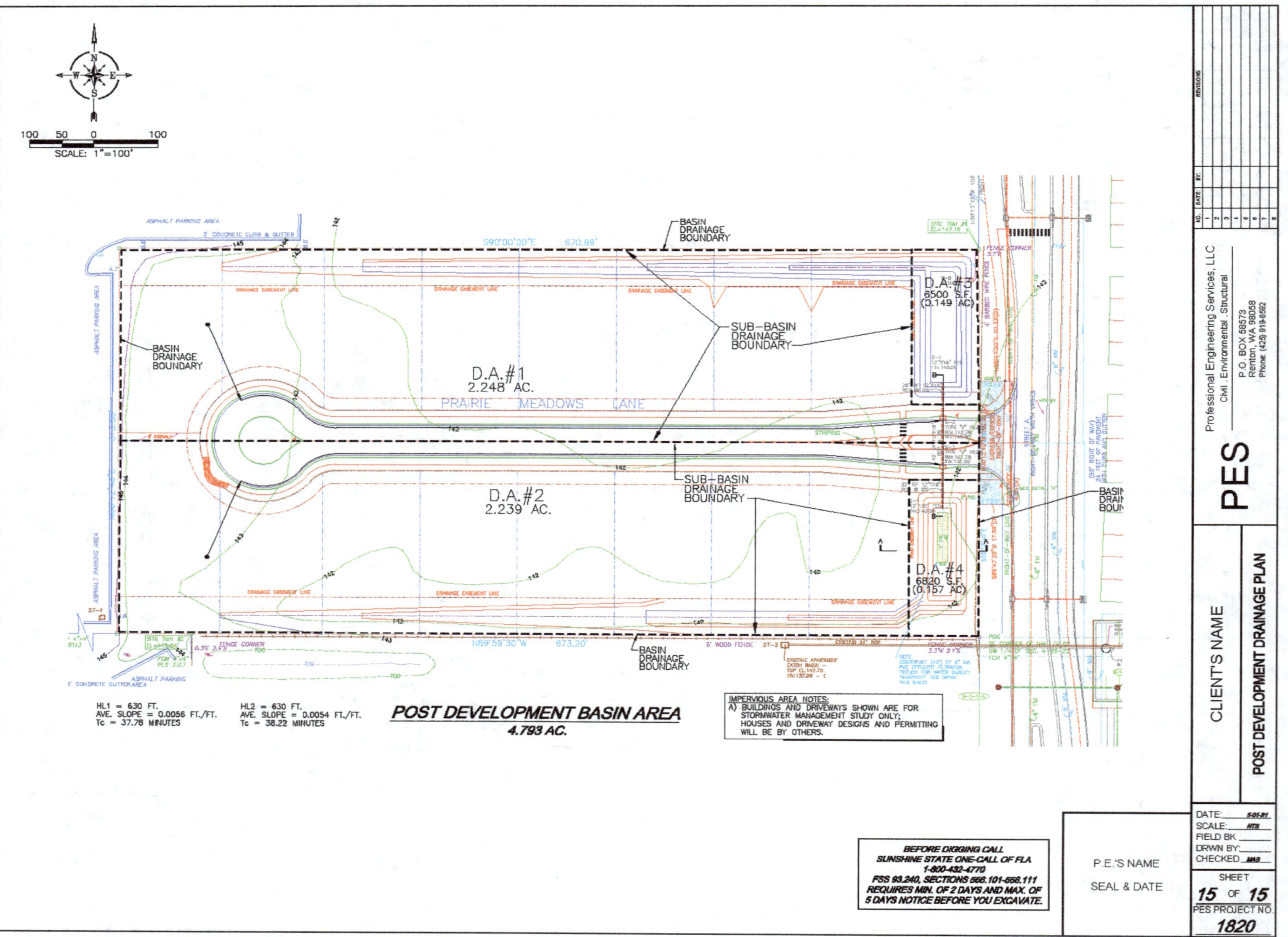
SCALE: 1"=100'
N E S W
ASPHALT PARKING AREA
2' CONCRETE CURB & GUTTER
BASIN DRAINAGE BOUNDARY
BASIN DRAINAGE BOUNDARY
S90°00'00"E   670.88'
DRAINAGE EASEMENT LINE
SUB-BASIN DRAINAGE BOUNDARY
D.A. #3
6500 S.F.
(0.149 AC.)
D.A. #1
2.248 AC.
PRAIRIE   MEADOWS   LANE
SUB-BASIN DRAINAGE BOUNDARY
D.A. #2
2.239 AC.
D.A. #4
6820 S.F.
(0.157 AC.)
N89°59'30"W   673.20'
BASIN DRAINAGE BOUNDARY
6' WOOD FENCE
BASIN DRAINAGE BOUNDARY
IMPERVIOUS AREA NOTES:
A) BUILDINGS AND DRIVEWAYS SHOWN ARE FOR
STORMWATER MANAGEMENT STUDY ONLY;
HOUSES AND DRIVEWAY DESIGNS AND PERMITTING
WILL BE BY OTHERS.
HL1 = 630 FT.
AVE. SLOPE = 0.0056 FT./FT.
Tc = 37.76 MINUTES
HL2 = 630 FT.
AVE. SLOPE = 0.0054 FT./FT.
Tc = 38.22 MINUTES
POST DEVELOPMENT BASIN AREA
4.793 AC.
BEFORE DIGGING CALL
SUNSHINE STATE ONE-CALL OF FLA
1-800-432-4770
FSS 93.240, SECTIONS 556.101-556.111
REQUIRES MIN. OF 2 DAYS AND MAX. OF
5 DAYS NOTICE BEFORE YOU EXCAVATE.
P.E.'S NAME
SEAL & DATE
156
ALI SHASTI, P.E.
PES
Professional Engineering Services, LLC
Civil . Environmental . Structural
P.O. BOX 68573
Renton, WA 98058
Phone (425) 919-8582
CLIENT'S NAME
POST DEVELOPMENT DRAINAGE PLAN
DATE:
SCALE: NTS
FIELD BK.
DRWN BY:
CHECKED
SHEET
15 OF 15
PES PROJECT NO.
1820

Project 3: Stonewood Place

8.1. Introduction

This property belonged to one of Access Engineering and Consulting Inc.'s client. It is a 26.41-acre property located in Pasco County, Florida, not relatively high but dry. It is not in a hundred-year flood zone; however, two small wetlands areas are located northeast. The site was zoned SFR. The main difference with Project 2 is that this property is not close to the public sanitary sewer system or potable water. Therefore, this project is designed with a future septic system and water well in mind for each lot.

A minimum of half-acre dry land is required for a property to be on the septic system. This requirement depends on the location of the property and the AHJ requirements. In this case, since both public water and sewerage are not available, each lot is designed to have a minimum of one-acre dry land. Also, since the site is rural, the zoning requires a minimum of one-acre lot size.

Curb and gutter, and storm sewer systems are provided to handle stormwater runoff in an urban setting. Storm sewer system directs the runoff to the appropriate facilities designed for this purpose. Dry or wet detention ponds, retention ponds (less common), stormwater vaults are examples of receiving storage facilities.

In a rural setting with inexpensive land, swale and ditches are provided to convey the stormwater runoff to the storage facilities and discharge location. The swale and ditch also act as low impact development (LID) with the capacity and capability to infiltrate storm runoff to the ground and some to evaporate. Occasionally, the professional engineer or designer may choose to design curb and gutter with the storm sewer system in a rural setting. After all, it is not practical or economically feasible to develop a rural subdivision in the middle of a big city or design an urban subdivision with plenty of inexpensive lands. In this case, and due to the site's steep slopes and sandy/loose and erodible soil, it was determined to design the stormwater management system using curb, gutter, and storm pipes to manage stormwater runoff. Either way, the requirements are often per the AHJs over the project area.

Coordinating with the AHJ for any future septic system and potable water for these projects is essential. It is important to ensure that the County Department of Health (DOH) allow an onsite sewer facility. The Southwest Florida Water Management District (SWFWMD) should also be contacted to obtain permission for any private water well and/or privately owned and maintained public water well is permitted.

8.2. Site Data

Here is a summary of the site data:

- Location — Pasco County, Florida
- Dimensions — 893 +/- feet wide x 1268 +/- feet long
- Project area — 26.41 acres (+/-)
- FEMA — Flood zone "X"
- Zoned — AR (Single Family Residential Rural)
- Future land use — RES-1
- Access to public street — Yes
- Topography — Slope from the southwest corner of the property and sheet flow to the northeast corner toward to the existing wetlands
- Critical areas — Two small wetlands 0.105 acre and 0.031 acre, and no other environmental natural features
- Trees — Existing orange grove with no significant trees
- Dry land — 26.27 acres +/-
- Potable water — Private water well
- Sanitary sewer — Private septic system

8.3. AR (Single Family Residential Rural)

Every jurisdiction has its zoning codes, which establish minimum requirements to abide by for laying out a site. Therefore, one of the early criteria to consider should be the zoning district where the project is and the minimum requirements and restrictions for that zoning district.

The following summarizes the minimum requirements for zone AR, where this project was located.

• Minimum lot size	One acre (dry land)
• Minimum ROW width	50 feet
• Minimum building setbacks	50 feet front
	50 feet rear
	25 feet sides

8.4. Stormwater Design Parameters

It is proposed to construct approximately 2,400 feet of paved roadway to serve nineteen proposed lots of 1.00 acre or more, plus a community park. It is assumed each lot will build around 6,000 SF of a residential home to include related impervious areas for driveway, sidewalk, swimming pool, spa, etc.

The EOR design and construct curbs and gutter, inlets, and pipes to collect and convey water runoff to the proposed dry detention pond on the northeast corner of the property. Control structures will also be provided, meeting the SWFWMD and Pasco County water quality and quantity requirements.

As in pre-development site conditions, the final discharge point of the site will not be altered. The proposed pond will overflow into the existing wetlands after meeting theAHJ standards and requirements.

As mentioned earlier, designing the stormwater management system is perhaps the most challenging part of preparing a site engineering and construction plan. It consumes most property and requires meeting or exceeding multiple agencies' rules and regulations. That is why it is crucial for the engineer and the designer to understand the requirements related to the property before diving deep into the design.

The following stormwater management design criteria are defined for this project,

1) Water Quality Treatment (Dry Detention Pond)

This project was designed as a dry detention pond, since the pond bottom was set at 203 feet, more than two feet above the SHWE at 201.70 feet. In other words, soil investigation revealed that the SHWE was well below the ground surface. This is not unusual for an orange grove, since orange trees need well-drained sandy and dry soil to survive. The water quality requirement for this project was to provide treatment for online storage volume based on a half-inch rainfall over the project area. Treatment is provided by percolating the half-inch volume through the bottom of the proposed pond.

2) Open vs. Closed Drainage Basin

The site is in a closed drainage basin. The Southwest Florida Water Management District (SWFWMD) defines a closed drainage basin as "a watershed in which the runoff does not have a surface outfall up to and including the 100-year flood level". Meaning that our project does not have a positive outfall, and the stormwater runoff discharges from the site after it reaches the hundred-year flood level.

In this case, the proposed detention pond must be designed to store runoff volume from the post-development minus the pre-development runoff volume before discharging off site. This requirement is in addition to the water quality requirement mentioned above and the below-stated peak discharge requirement.

3) Water Quantity Requirement

The requirement is to meet or exceed the twenty-five-year, twenty-four-hour peak discharge. In other words, we need to ensure storm runoff release from the site during post-development is equal to or preferably blows the pre-development peak discharge for a twenty-five-year, twenty-four-hour storm event for our specific location.

8.5. Urban vs Rural Subdivisions

To save time and space, I focus only on this project's unique characteristics and elements compared to the previous Project 2. In addition, partial construction plans for this project will be presented here and not the complete set.

    1) Please note that the zoning of this site is a single family residential plus rural, which means that the site is in a rural setting and not in an urban environment. The main difference is in the density or minimum required size of each lot to accommodate private water wells and septic systems. If potable water was available to the site, the lot size could have been reduced to a half-acre if approved by the AHJ. The Department of Health has a minimum required distance between the proposed water

well and the drain field area of a septic system that must be met to obtain their permit. Therefore, a much larger land area must be allocated for each lot to manage water and wastewater utilities.

2) There is a small area of wetland on-site that must be preserved. On occasion, one might encroach the buffer or the wetland area, but the impact needs to be mitigated. The mitigation process and the follow-up monitoring of the mitigation area can be time-consuming, laborious, and expensive, assuming it is successful after implementation. Therefore, it is crucial to delineate the buffer area during the design and follow up construction and stay away from it as much as possible.

3) The minimum ROW width was fifty feet in the previous Project 2, Chapter 7, and it is fifty feet for this project. Although this project is zoned rural, using a curb, gutter, and storm sewer system allows a fifty-foot ROW. Otherwise, a much wider ROW width is required to accommodate a rural section with ditches.

4) Also, notice that the setback distances are longer in a rural lot than in an urban lot.

Please be advised that none of the above criteria are necessarily negative or positive points against or for a project site. The land cost typically is less expensive in rural areas than urban areas. But the cost of construction is higher in a metropolitan area than in a rural setting. That is why performing due diligence and evaluating a site is vital in acquiring any project site, as described in Chapter 1, Site Evaluation.

8.6. Final/Partial Construction Plans

The following pages explain and include partial construction plan sheets 1, 3, 7, 17, and 18 of 18.

1)   Sheet 1 of 18 - Cover

This sheet is very similar to the sheet presented in Chapter 7, Project 2. We try to avoid repetition to cover what has already been explained. One important note is that the cover sheet may only include the project location for much larger projects. For large roadway or transit-oriented projects, the cover sheet may indicate the beginning and end of the construction route for the project with the associated notes, key sheets layout, etc. One or more sheets may cover the index of sheets in those scenarios, and one or more sheets to cover the legend. The general notes and specifications and some details typically will be included in a specification report.

2) Sheet 3 of 18 - Site Plan

This site plan includes configuration and dimensions of the proposed lots, roadway and entrance, tracts, and easements, proposed pond locations and sizes, access easements, etc. It also shows a proposed park.

It also provides sufficient geometric dimensions for structures envelopes, driveways, roadways, curb and gutter, sidewalk, cul-de-sac, landscape islands, and ponds for the surveyor and the contractor to perform the field's horizontal layout.

This sheet also provides building setbacks (BSBL) lines and dimensions from the ROW and the property lines. It gives the areas in square footage and acreage for each lot.

There are six tracts and ten easements shown on this sheet for this project. The following paragraphs provide a brief explanation of each tract or easement along with their intended use,

1) Tract "A": Tract A includes the entire proposed roadway ROW within the subdivision. Some agencies may allow easements in place of a designated tract. The difference is if an easement is permitted, then the total area of each lot may be allowed to use half of the roadway (to the centerline) area as the lot area to meet the minimum lot requirement. It usually is not permitted with a tract. The tract is a separate legal parcel. As one can see, each of the lots meets the 1.0-acre minimum requirement in this project. For more extensive and complicated projects, one may seek the advice of an experienced real estate attorney since the concept of the easement vs. tract may not be as simple as it appears to be.

2) Tract "B": Tract B is a park set aside for the development and those living in the subdivision. The park is a zoning requirement by the AHJ based on the subdivided parcel size, number of lots, lot sizes, etc. A note provided states no impervious area is allowed on this tract. Playground equipment is most likely allowed in this area without creating any sizeable, paved area the HOA approves.

3) Tract "C," "C-1," "D," and "E": These tracts are designated areas for the proposed or future ROW that are outside of the subdivision. They will be dedicated to the county, and the agency will have the right and resources to improve these rights-of-way as they see fit.

4) Easement "A," "A-1," "A-2," "B," "C," "D," "E," and "H": These easements are designated areas for the drainage or access to the drainage easements (i.e., ponds, swales) for routine maintenance, repair, etc.

5) Easement "F" and "G": These easements are designated areas for 20 feet landscape buffer required by the County zoning code.

3)   Sheet 7 of 18 – Paving, Grading and Drainage Plan (PG & D)

This sheet may be the more challenging in the entire set to compute correctly for an optimum design. It should have sufficient elevations to define the proposed site topography and contours. It should include structures envelopes, corner elevations, driveways, roadways, curbs and gutters, sidewalks, cul-de-sacs, landscape islands, and ponds for the surveyor and the contractor to perform the field's horizontal layout.

Let's compare this PG & D sheet with the one presented in Chapter 7 for Project 2. It is easy to see that many more spot elevations are provided on this sheet. The elevation difference on this site from the southwest corner of the property to the northeast is approximately twenty-three feet, while this difference for Project 2 is about two feet.

Many more spot elevations are required to balance the excavation (cut and fill) volume to grade this site accurately. Another important reason is additional spot elevations for the contractor or the surveyor to layout the proposed site in the field.

The goal is to guide the storm runoff toward the proposed low area (i.e., swales, ditches, and ponds) and away from the building pads. The roadway grading is such that the storm runoff sheet from the centerline (crown) of the roadway to the proposed gutters, then to the inlets located on the low spots, and eventually discharges to the ponds via the proposed storm sewer system. The proposed ponds are designed to detain water to the specified design stage (hundred-year storm event for this project) and slowly discharge to the discharge point—a wetland, in this case.

It is important to remember that, depending on the quality of the wetland, the discharge should not adversely impact the hydroperiod elevation in it.

Two crucial soil profile data indicated on this sheet provide the actual soil properties at the pond locations and the seasonal high-water table (SHWT) at each location. The soil profiles suggest that the seasonal high groundwater elevation is about thirteen feet below the existing ground elevation and about five feet below the proposed bottom of the pond elevation. This depth and the type of material for the sandy soil make this site a good candidate for the dry pond.

For a dry pond to function as a dry pond, the SHWT should be located a minimum of two feet below the proposed bottom of the pond elevation. The existing ground elevations at each B-01 and B-02 were about 208 feet. Geotechnical engineering estimated the SHWT about thirteen feet below the grade. It means the seasonal high-water table was estimated to be at 195 feet. We set the bottom of the proposed pond at an elevation of 203 feet, which is about eight feet above the SHWE, and provides ample room for the possible groundwater fluctuations.

4)   Sheet 17 of 18 - Pre-Development Drainage Plan

The primary purpose is to provide predevelopment drainage data to facilitate drainage computation and report. It is not a construction plan sheet and does not need to be included in construction plans offered to contractors for bidding, except for information only. However, it is an integral part of the drainage report submitted to permitting agencies reviewing and approving stormwater management design.

Let us review some of the information provided on this sheet.

It is essential to mention here that the predevelopment drainage plan can utilize the actual survey sheet with all existing features. Drainage basin boundary will be delineated on this sheet. The area within the basin drainage boundary is the total predevelopment basin area. Comparing this site to Project 2, Chapter 7, which was relatively flat, has an elevation drop of approximately twenty-three feet from the south to the north side of the property. The stormwater runoff flows from the southwest corner toward the northeast corner to an existing depressed or wetland area. A small portion of the property on the south side flows to the south to the existing street ROW. Therefore, it is necessary to separate that portion of the site as a separate drainage sub-basin area (Area #1) and Area #2, as shown on the predevelopment drainage plan.

The flow length, average slope, and time of concentration are also shown on this sheet. The computation for the concentration-time followed the process covered in Chapter 7 for Project 2 and tried to avoid repetition here.

5) Sheet 18 of 18 - Post-Development Drainage Plan

The primary purpose is to provide post-development drainage data to facilitate drainage computation and report. It is not a construction plan sheet and does not need to be included in construction plans offered to contractors for bidding, except for information only. However, it is included in the drainage report submitted to permitting agencies reviewing and approving stormwater management design.

In addition to delineating the basin drainage boundaries, it is essential to subdivide the entire drainage area into smaller sub-basin drainage boundaries to represent the actual storm runoff behavior in the field and facilitate computations.

As mentioned earlier, the predevelopment drainage basin for the existing site conditions was sub-divided into two drainage basins. The post-development drainage basin has been sub-divided into fourteen sub-basins. The intent is to cover all possible drainage flow through the site, around and between the buildings, through the streets and curb and gutters, through the storm sewer systems, and finally discharging to the proposed pond. The proposed pond is wrapped around the existing wetland to utilize the current low area and, at the same time, optimize the site for the most usable lots.

In addition to basin boundaries and other information related to stormwater management system data, a note is provided on this sheet about the proposed homes. It is unknown what size home and other related impervious surfaces, such as a swimming pool and driveway, will be built for each lot. Therefore, the stormwater management report assumed that about 4,000 SF of home and other related impervious areas would be constructed on each lot to finalize the drainage report. This assumption is conservative since most average homes are below this threshold, particularly two-story SFR. If the total impervious area draining to the pond stays below the assumed number, there is no need to modify the drainage report and the permit approval by the AHJ.

If, for whatever reason, the total impervious area exceeds the assumed number, then the AHJ will require a modification to the permit to justify the additional impervious area . This modification may be submitted as part of the final as-built certification to the AHJ.

The following pages represent the partial construction plans of Sheets 1, 3, 7, 17, and 18 of 18.

GENERAL NOTES AND SPECIFICATIONS

1. Unless otherwise noted on the plans, all construction methods, workmanship, materials, and testing shall conform to the "Florida Department of Transportation (FDOT)'s "Standard Specification for Road and Bridge Construction." Its current supplements, standard index drawings, the county, or any other applicable agency's rules and regulations.
2. The Contractor shall perform survey layout and testing.
3. The Contractor must coordinate all work within the right-of-way with the city, county, or FDOT and restore all disturbed areas within the right-of-way to the original condition or better.
4. Locations, elevations, and dimensions of existing utilities, structures, and other features are shown according to the best information available at the time of preparation of these plans. The Contractor shall verify all information affecting this work before construction.
5. The Contractor shall use extreme caution in areas of buried utilities and shall provide at least 48 hours' notice to the utility companies before construction to obtain field locations of existing underground utilities. The Contractor shall protect and maintain the operation of existing utilities during construction.
6. The Contractor shall check plans for conflicts and discrepancies before construction. The Contractor shall notify the engineer of any disputes before performing any work in the affected area.
7. The Contractor must notify the Engineer of Record where existing utility services interfere with the proposed construction.
8. The Contractor is responsible for repairing any damage to existing facilities, above or below ground, that may occur due to work performed by the Contractor.
9. All underground utilities must be in place and tested or inspected before the base and surface construction.
10. Utility installation shall conform to the requirements of the applicable federal, state, and local regulations, including but not limited to the Florida Department of Environmental Protection (FDEP), Department of Health (DOH), the Occupational Safety and Health Administration (OSHA) Standard.
11. The Contractor's responsibility is to become familiar with the permit and inspection requirements specified by the various governmental agencies and the Engineer of Record. The Contractor shall obtain all necessary permits before construction and schedule inspections according to agency instructions. All work performed shall comply with the regulations and ordinances of the various governmental agencies having jurisdiction over the project.
12. At least three (3) working days before construction, the Contractor shall notify the engineer and appropriate agencies and supply their name, starting date, projected construction schedule, all required shop drawings, and other information as needed. Any work performed before proper notifications may be subject to the removal and replacement at the Contractor's expense.
13. The site shall be initially graded such that no offsite area will be adversely affected by stormwater runoff. The Contractor shall minimize the stormwater Runoff discharging to the adjacent right-of-way. The existing drainage facilities within the right-of-way shall be protected using hay bales, temporary grading, revegetation, etc. Any disturbed area within the right-of-way shall be sodded. The Contractor shall final grade and hand rake all exposed earthwork areas immediately before installing sod. All disturbed areas shall be protected using proper erosion and sediment control devices from erosion caused by stormwater runoff.
14. Upon completion of construction activities, the stormwater management system shall be cleaned of oil debris, silts, lime rock, and foreign materials.
15. The Contractor shall verify minimum separation between water and sanitary sewer systems (18" minimum vertical and 10' minimum horizontal). Upon completion of construction, the Contractor shall provide written verification documenting the minimum separation as constructed.
16. Typical minimum cover over utilities shall be 36".
17. The Contractor must enforce all safety regulations during the construction and maintenance of this project. The Contractor or their representative shall be responsible for the control and safety of the traveling public and the safety of their personnel.
18. The Contractor shall comply with all Federal/State Occupational Safety and Health Act (OSHA) Standards and any other rules and regulations applicable to construction and or maintenance activities in the State of Florida. The Contractor shall also comply with Chapter 442, Florida Statutes (Toxic Substances in the Workplace), and any CITY, COUNTY, or any other applicable agency's rules and regulations regarding safety.
19. All trenching and excavation operations must conform to the new "Trench Safety Act" requirements incorporated into OSHA standards 29 CFR 1926

Subpart P final rule and the State of Florida Trench Safety Act. The Contractor shall recognize the OSHA safety standards, agree to abide by them, and identify the compliance cost. Further Contractor operations shall comply with 29CFR 1910.252 and NFPA 51B for cutting and welding procedures.

20. All persons on (CITY, COUNTY OR STATE) property and in the area where the noise level exceeds 85db must wear hearing protection that complies with ANSI S3.19-74 (earmuffs and or approved earplugs). This requirement includes areas where noisy equipment is used (i.e., jackhammers, electric or air drills, heavy equipment with open cabs, pipe cutting saws, etc.) and posted in the plant.
21. All work conducted in an elevated position shall comply with 29CFR 1910.269.
22. All traffic control markings and devices shall conform to the "Manual on Uniform Traffic Control Devices" provision prepared by the U.S. Department of Transportation, Federal Highway Administration.
23. All design applications, installation, maintenance, warning devices, and barriers necessary to protect the public and the workmen within the project shall be per the minimum standards outlined in the manual on Uniform Traffic Control Devices (MUTCD).
24. All Contractor-owned or controlled vehicles and or equipment operated on or within ten (10) feet of the roadway shall be equipped with a minimum of one amber 360 degrees Class I warning device. This device must meet minimum standards for utility construction purposes, such as a minimum of 500,000 candlepower and visible from 360 degrees of mounting. The warning device(s) must be always in operation that a vehicle/equipment is on the roadway or within the ten (10) feet of the right-of-way area and not in a "normal" travel status.
25. Area designated by the Owner shall be utilized for equipment storage and covered material staging. Construction materials and supplies shall be stored only in those designated areas approved by the Owner.
26. Regarding the inspection of the project, the Southwest Florida Water Management District (SWFWMD) and the COUNTY inspectors shall have the same right of review and assessment as the Engineer of Record. However, it is appropriate for the Engineer of Record to resolve questions, difficulties, and disputes regarding interpretation of contract documents or construction; SWFWMD and the COUNTY reserve the right of final concurrence on such matters.
27. The Engineer's approval shall be obtained on all proposed field changes, alterations, or additions to the approved plans by the Contractor before implementing such modifications, alterations, or additions. The AHJ's Inspector shall be notified of all such changes. Upon completion of construction, the Owner will require one set of complete "As-Built." Contractor to supply certified "As-Built" to the Engineer of Record.
28. The Contractor shall obtain a separate permit for any on-site burning.
29. If any evidence of historic resources during construction activities, including aboriginal or historic pottery, prehistoric stone tools, bone or shell tools, historic trash pits, or historic building foundation, is discovered, work shall immediately stop. The Florida Department of Historical Resources (State Historic Preservation Officer) and the County shall be notified within two working days of the resources found on the site.
30. All proposed signs must be applied for, approved, and permitted individually, apart from the ultimately approved site engineering plan. Approval of the site plan does not constitute approval of any signage.
31. All parking and accessible spaces will be appropriately signed and striped per Florida Statute 316, the Manual on Uniform Traffic Control Devices (MUTCD), or other applicable standards.
32. The Engineer of Record certifies that the site has been designed following the American Disability Act.
33. Parking spaces, directional arrows, and stop bars shall be striped in white. It shall be the Owner/Contractor's responsibility to install the sign and stripe the site following applicable standards correctly.
34. All clear-site areas shall be kept free of any signage plantings, trees, etc., over three and one-half feet in height.
35. All structures, including buffer walls, retaining walls, fencing, signage, etc., may require a separate building permit.
36. No irrigation system or landscaping shall be installed in any county or state right of way without the appropriate right of way permit.
37. The owner/developer acknowledges that the site and subsequent building permits shall comply with all rezoning conditions.
38. The proposed roadway system is in substantial conformance with the Manual of Uniform Minimum Standards for Design, Construction, and Maintenance of Streets and Highways, State of Florida, when these plans were prepared.

PRELIMINARY / CONSTRUCTION PLANS
FOR
STONEWOOD PLACE

VICINITY MAP
PASCO COUNTY ATLAS MAP 30
SECTION 25, TOWNSHIP 24 SOUTH, RANGE 20 EAST

36	31	32	33	34	35	36	31
1	6	5	4	3	2	1	6
12	7	8	9	10	11	12	7
13	18	17	16	15	14	13	18
24	19	20	21	22	23	24	19
25	30	29	28	27	26	25	30
36	31	32	33	34	35	36	31
1	6	5	4	3	2	1	6

CLIENT: CLIENT'S NAME / STREET ADDRESS / CITY, STATE ZIP CODE / PHONE NUMBER

SURVEYOR: SURVEYOR'S NAME / STREET ADDRESS / CITE, STATE ZIP CODE / PHONE NUMBER

CONTRACTOR: TO BE DETERMINED

PROPOSED MINIMUM DEED RESTRICTIONS

1. A definition for the term "surface water management system facilities" substantially as follows:
   The surface water management system facilities shall include, but are not limited to: all inlets, ditches, swales, culverts, water control structures, retention and detention areas, ponds, lakes, floodplain compensation areas, wetlands and any associated buffer areas, and wetland mitigation areas.
2. No construction activities may be conducted relative to any portion of the surface water management system facilities. Prohibited activities include, but are not limited to: digging or excavation; depositing fill, debris or any other material or item; constructing or altering any water control structure; or any other construction to modify the surface water management system facilities. If the project includes a wetland mitigation area, as defined in section 1.7.24, or a wet detention pond, no vegetation in these areas shall be removed, cut, trimmed or sprayed with herbicide without specific written approval from the District. Construction and maintenance activities which are consistent with the design and permit conditions approved by the District in the Environmental Resource Permit may be conducted without specific written approval from the District.
3. An amendment of the declaration of protective covenants, deed, restrictions or declaration of condominium affecting the surface water management system facilities shall have the prior written approval of the District.
4. The restrictions shall be in effect for at least 25 years with automatic renewal periods thereafter.
5. The surface water management system facilities are located on land that is designated common property on the plat or are located on land that is subject to an easement in favor of all of the lot owners within the subdivision.
6. The permittee shall be responsible for operation and maintenance of the surface water management system facilities until the first successful reinspection conducted pursuant to the Environmental Resource Permit. The transfer of responsibility to the lot owners will not be effective until this District approves the transfer in writing.
7. The lot owners shall be jointly and severally responsible for operation and maintenance of the surface water management system facilities after the first successful reinspection.
8. Operation and maintenance, and reinspection reporting shall be performed in accordance with the terms and conditions of the Environmental Resource Permit.
9. The District has the right to take enforcement measures, including a civil action for injunction and/or penalties, against any lot owner(s) to compel such lot owner(s) to correct any outstanding maintenance problems with the surface water management system facilities.

INDEX OF SHEETS

SHEET NO.	SHEET DESCRIPTION
1.	COVER SHEET AND GENERAL NOTES
2.	SURVEY PLAN (EXISTING SITE CONDITIONS)
3.	PRELIMINARY SITE PLAN
4.	SITE DATA
5.	RIGHT-OF-WAY, EASEMENT AND TRACT PLAN
6.	TRAFFIC SIGNAGE AND MARKING PLAN
7.	PAVING, GRADING AND DRAINAGE PLAN
8.	STONECREST TRAIL - PLAN AND PROFILE
9.	WINDWOOD LOOP - PLAN AND PROFILE
10.	ROCKMERE COURT - PLAN AND PROFILE
11.	CROSS SECTIONS
12.	LANDSCAPE PLAN
13.	LANDSCAPE SPECIFICATIONS
14.	GENERAL DETAILS
15.	BEST MANAGEMENT PLAN
16.	BEST MANAGEMENT DETAILS
17.	PRE-DEVELOPMENT DRAINAGE PLAN
18.	POST DEVELOPMENT DRAINAGE PLAN

ALI SHASTI, P.E.

PES
Professional Engineering Services, LLC
Civil . Environmental . Structural
P.O. BOX 58573
Renton, WA 98058
Phone: (425) 919-8692

STONEWOOD PLACE
COVER SHEET AND GENERAL NOTES

DATE:	5.01.31
SCALE:	NTS
FIELD BK	
DRWN BY	
CHECKED	MB
SHEET	1 OF 18
PES PROJECT NO.	0610

P.E.'S NAME
SEAL & DATE

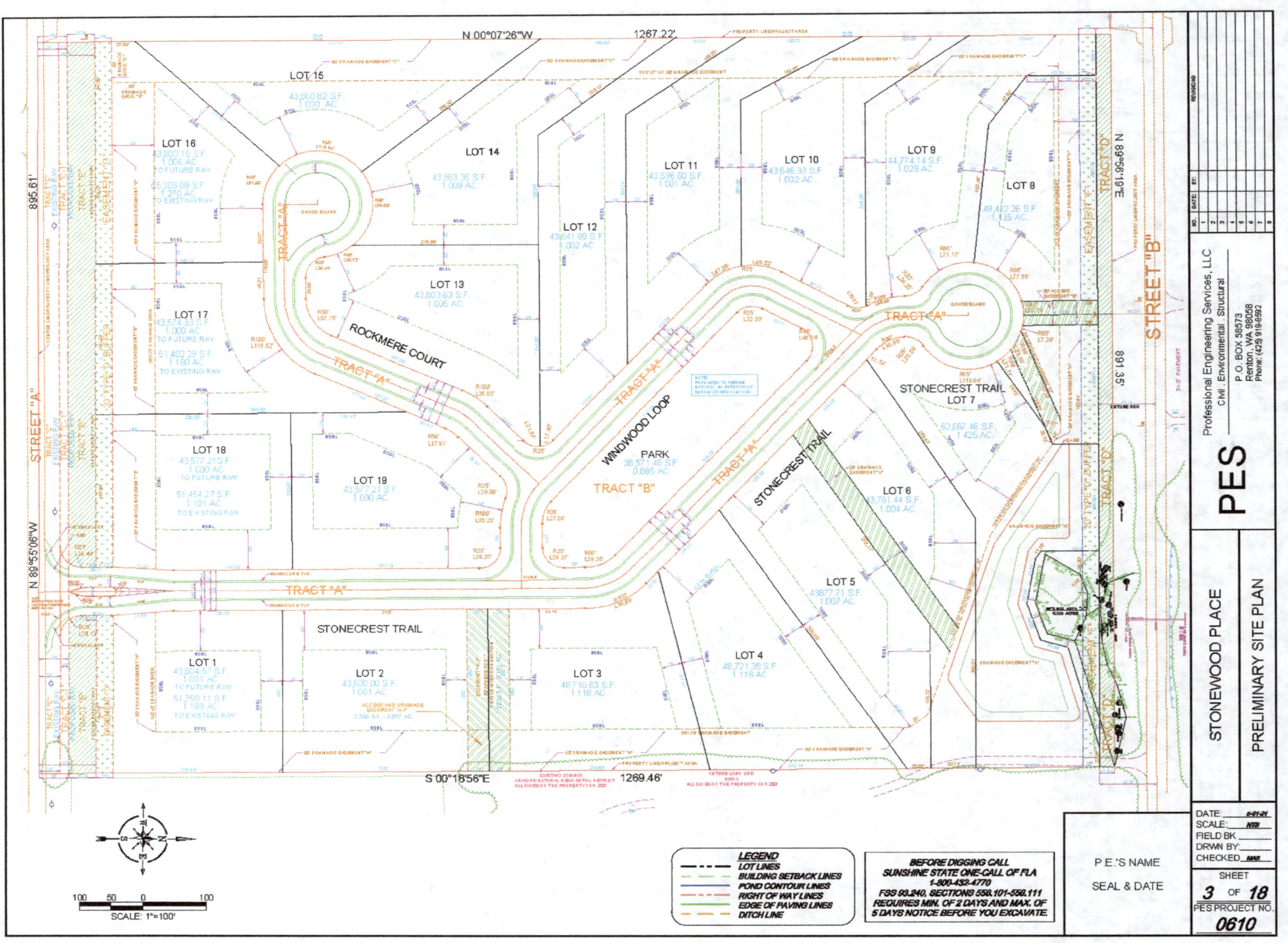
SITE DESIGN & ENGINEERING
PES
Professional Engineering Services, LLC
Civil, Environmental, Structural
P.O. BOX 58573
Renton, WA 98058
Phone: (425) 919-8592
STONEWOOD PLACE
PRELIMINARY SITE PLAN
DATE:
SCALE: NTS
FIELD BK
DRWN BY:
CHECKED
P.E.'S NAME
SEAL & DATE
SHEET
3 OF 18
PES PROJECT NO. 0610

LEGEND
LOT LINES
BUILDING SETBACK LINES
POND CONTOUR LINES
RIGHT OF WAY LINES
EDGE OF PAVING LINES
DITCH LINE

BEFORE DIGGING CALL
SUNSHINE STATE ONE-CALL OF FLA
1-800-432-4770
FSS 93.240, SECTIONS 550.101-550.111
REQUIRES MIN. OF 2 DAYS AND MAX. OF
5 DAYS NOTICE BEFORE YOU EXCAVATE

SCALE: 1"=100'
100 50 0 100

STREET "B"
STREET "A"
N 00°07'26"W 1267.22'
S 00°18'56"E 1269.46'
N 89°55'06"W 895.61'
N 89°56'19"E 891.35'

TRACT "A"
TRACT "B"
TRACT "D"
EASEMENT "F"
ROCKMERE COURT
WINDWOOD LOOP
STONECREST TRAIL
PARK

LOT 1
LOT 2
LOT 3
LOT 4
LOT 5
LOT 6
LOT 7
LOT 8
LOT 9
LOT 10
LOT 11
LOT 12
LOT 13
LOT 14
LOT 15
LOT 16
LOT 17
LOT 18
LOT 19

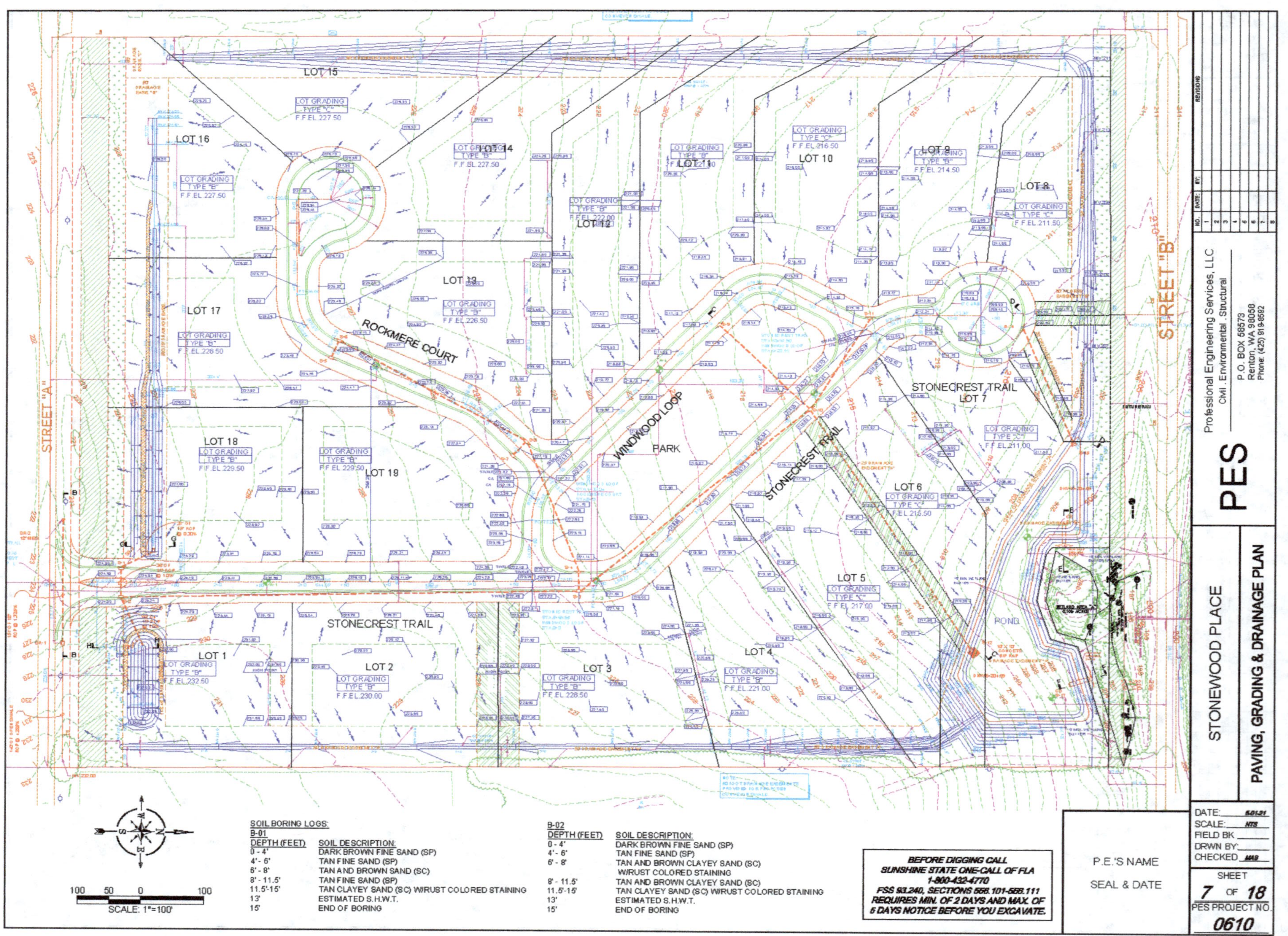
STREET "B"
STREET "A"
LOT 1
LOT 2
LOT 3
LOT 4
LOT 5
LOT 6
LOT 7
LOT 8
LOT 9
LOT 10
LOT 11
LOT 12
LOT 13
LOT 14
LOT 15
LOT 16
LOT 17
LOT 18
LOT 19
STONECREST TRAIL
ROCKMERE COURT
WINDWOOD LOOP
PARK
POND
LOT GRADING TYPE "C" F.F.EL. 227.50
LOT GRADING TYPE "B" F.F.EL. 227.50
LOT GRADING TYPE "B" F.F. EL. 227.50
LOT GRADING TYPE "B" F.F.EL. 222.00
LOT GRADING TYPE "B" F.F.EL. 226.50
LOT GRADING TYPE "B" F.F.EL. 226.50
LOT GRADING TYPE "C" F.F.EL. 216.50
LOT GRADING TYPE "B" F.F.EL. 214.50
LOT GRADING TYPE "B" F.F.EL. 211.50
LOT GRADING TYPE "C" F.F.EL. 211.00
LOT GRADING TYPE "C" F.F.EL. 215.50
LOT GRADING TYPE "C" F.F. EL. 217.00
LOT GRADING TYPE "B" F.F.EL. 229.50
LOT GRADING TYPE "B" F.F. EL. 229.50
LOT GRADING TYPE "B" F.F.EL. 232.50
LOT GRADING TYPE "B" F.F.EL. 230.00
LOT GRADING TYPE "B" F.F.EL. 228.50
LOT GRADING TYPE "B" F.F.EL. 221.00

SOIL BORING LOGS:
B-01
DEPTH (FEET)   SOIL DESCRIPTION:
0 - 4'   DARK BROWN FINE SAND (SP)
4' - 6'   TAN FINE SAND (SP)
6' - 8'   TAN AND BROWN SAND (SC)
8' - 11.5'   TAN FINE SAND (SP)
11.5'-15'   TAN CLAYEY SAND (SC) W/RUST COLORED STAINING
13'   ESTIMATED S.H.W.T.
15'   END OF BORING

B-02
DEPTH (FEET)   SOIL DESCRIPTION:
0 - 4'   DARK BROWN FINE SAND (SP)
4' - 6'   TAN FINE SAND (SP)
6' - 8'   TAN AND BROWN CLAYEY SAND (SC) W/RUST COLORED STAINING
8' - 11.5'   TAN AND BROWN CLAYEY SAND (SC)
11.5'-15'   TAN CLAYEY SAND (SC) W/RUST COLORED STAINING
13'   ESTIMATED S.H.W.T.
15'   END OF BORING

SCALE: 1"=100'
100   50   0   100

BEFORE DIGGING CALL
SUNSHINE STATE ONE-CALL OF FLA
1-800-432-4770
FSS 93.340, SECTIONS 556.101-556.111
REQUIRES MIN. OF 2 DAYS AND MAX. OF
5 DAYS NOTICE BEFORE YOU EXCAVATE.

Professional Engineering Services, LLC
Civil. Environmental. Structural
P.O. BOX 68573
Renton, WA 98058
Phone: (425) 919-8592
PES

STONEWOOD PLACE
PAVING, GRADING & DRAINAGE PLAN

DATE:
SCALE: NTS
FIELD BK
DRWN BY:
CHECKED MAB
P.E.'S NAME
SEAL & DATE
SHEET
7 OF 18
PES PROJECT NO.
0610
REVISIONS
NO. DATE: BY.

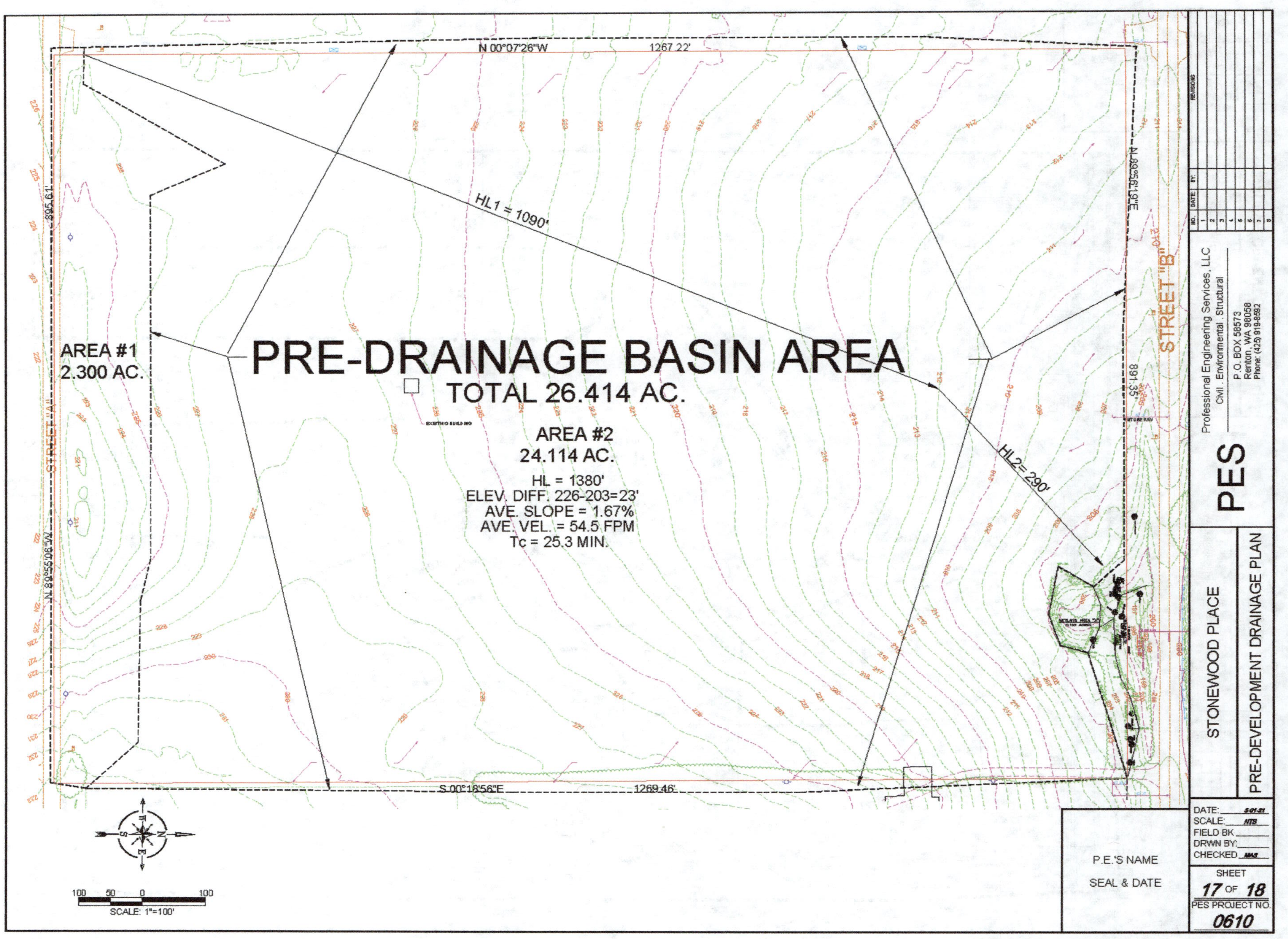

Professional Engineering Services, LLC
Civil, Environmental, Structural
P.O. BOX 58573
Renton, WA 98058
Phone: (425) 919-8592

PES

STONEWOOD PLACE

PRE-DEVELOPMENT DRAINAGE PLAN

DATE: 5-01-21
SCALE: NTS
FIELD BK
DRWN BY:
CHECKED MAS

P.E.'S NAME

SEAL & DATE

SHEET 17 OF 18
PES PROJECT NO.
0610

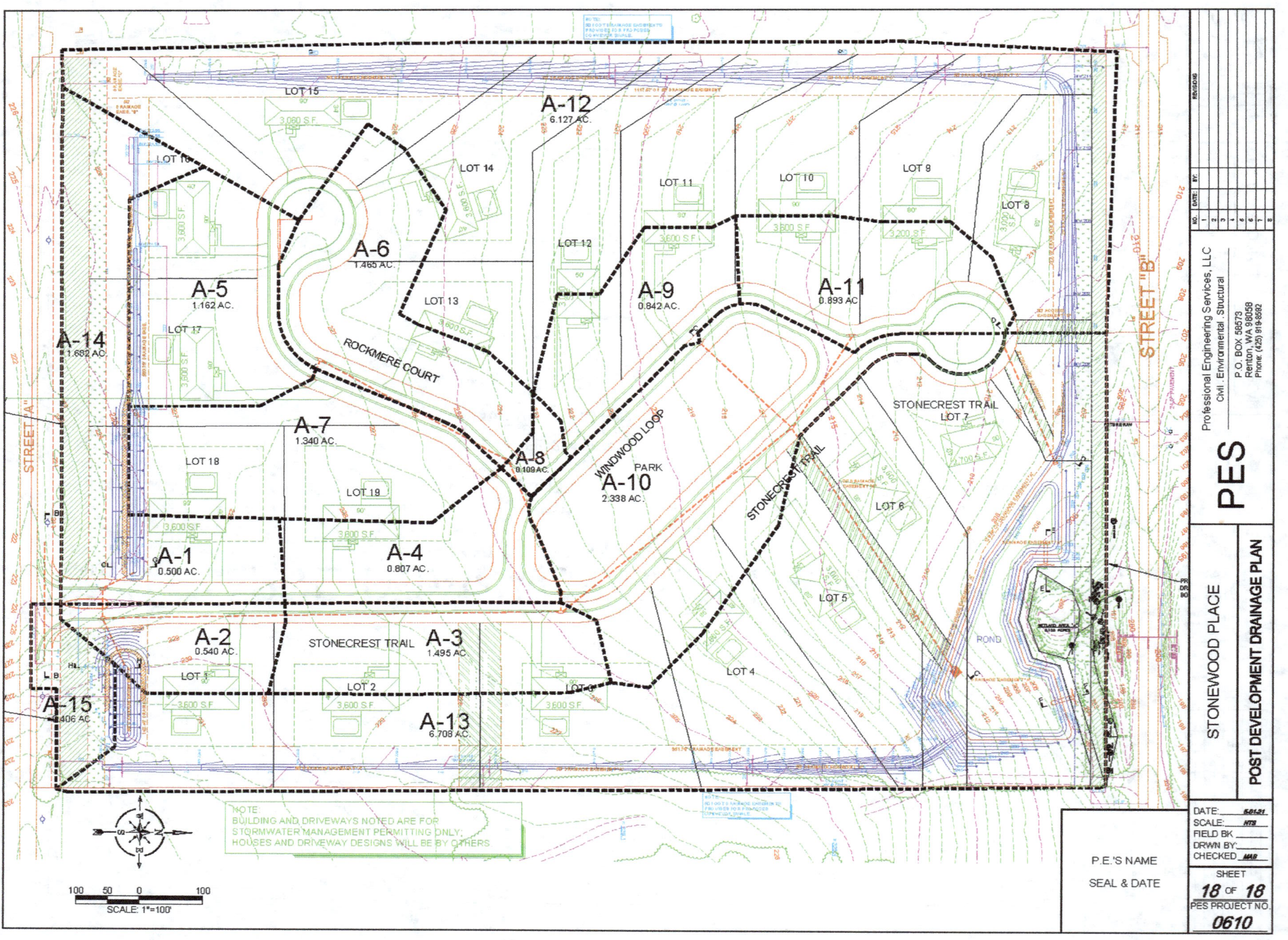
PES
Professional Engineering Services, LLC
Civil . Environmental . Structural
P.O. BOX 58573
Renton, WA 98058
Phone (425) 919-8582
STONEWOOD PLACE
POST DEVELOPMENT DRAINAGE PLAN
DATE:
SCALE: NTS
FIELD BK
DRWN BY:
CHECKED
SHEET 18 OF 18
PES PROJECT NO. 0610
P.E.'S NAME
SEAL & DATE
STREET "A"
STREET "B"
STONECREST TRAIL
WINDWOOD LOOP
ROCKMERE COURT
PARK
POND
NOTE:
BUILDING AND DRIVEWAYS NOTED ARE FOR
STORMWATER MANAGEMENT PERMITTING ONLY;
HOUSES AND DRIVEWAY DESIGNS WILL BE BY OTHERS.
LOT 1
LOT 2
LOT 3
LOT 4
LOT 5
LOT 6
LOT 7
LOT 8
LOT 9
LOT 10
LOT 11
LOT 12
LOT 13
LOT 14
LOT 15
LOT 16
LOT 17
LOT 18
LOT 19
A-1 0.500 AC.
A-2 0.540 AC.
A-3 1.495 AC.
A-4 0.807 AC.
A-5 1.162 AC.
A-6 1.465 AC.
A-7 1.340 AC.
A-8 0.109 AC.
A-9 0.842 AC.
A-10 2.338 AC.
A-11 0.893 AC.
A-12 6.127 AC.
A-13 6.708 AC.
A-14 1.682 AC.
A-15 0.406 AC.
SCALE: 1"=100'
N

CHAPTER 9: COMMERCIAL SITE

9.1. Introduction

This project is proposed to replace the existing mom-and-pop hotel. The AEC, Inc. was hired to design and replace the current two-story-old hotel with a new franchise hotel and prepare construction plans and permit documents for a proposed three-story chain hotel.

It was proposed to construct approximately 16,000 SF, three-story hotel plus a single-story storage building and required parking, driveway, sidewalk, swimming pool, and spa. A portion of the property in the front was left vacant for future improvements, preferably a chain dine-in restaurant. However, the stormwater management report assumed seventy percent of impervious areas accounted for the design of the proposed stormwater management system. If the total impervious area (buildings, driveway, parking, sidewalk, etc.) remained below seventy percent impervious surface, there would be no need to go through the stormwater management permitting from the water management district. Although, the local agencies, in this case, may review and approve the drainage plans and computations.

The site is a 3.36-acre property located in Highlands County, Florida, that is not exceptionally high but dry. It is not in a hundred-year flood zone. The site is zoned commercial (C2), allowing many uses, including hotels and dine-in restaurants.

9.2. Site Data

Here is a summary of the site data:

- Location Highlands County, Florida
- Property dimensions   258 feet wide x 567 feet long
- Area 3.36 acre (+/-)
- FEMA Not in FEMA identifies flood district
- Zoned C2 (commercial)
- Land use   Hotel and restaurant

- Access to public street Direct access to public road
- Topography Fairly flat slopping toward the back   (west side)
- Critical areas   No wetlands, steep slopes or water body or creeks
- Trees Small to medium-sized trees and   vegetations (No majestic oak or other significant trees)
- Potable water   Available
- Sanitary sewer   Available
- Power   Available

9.3. C2 (Commercial)

As mentioned earlier, jurisdictions establish minimum zoning requirements for laying out a site. Therefore, one of the early criteria should be the zoning district where the project is located and the minimum requirements and restrictions for that specific zoning district.

The following is a summary of the minimum requirements for zone C2 – commercial for this project:

- Minimum building setbacks   25 feet front
25 feet rear
10 feet sides *

The architectural and site design gets a little more complicated to design a hotel, restaurant, or any other facility part of a franchise. Like AHJ, each franchise has its own rules and regulation and, depending upon the facility, may be more restrictive than the AHJ. They may have additional requirements that have to be met. For example, many franchise hotels have their flags and require the site to provide a location to display flags in front of the hotel. They may want to enclose the dumpsters, which may or may not be required by the AHJ. They may also have more regular and designated accessible parking spaces. These are just some of the examples. Therefore, it is essential to get familiar with their requirement in addition to the AHJ requirements right off the back.

*NOTE: There are also landscape buffer requirements that must be met to obtain approval from the AHJ. This buffer relates to the adjacent properties' zoning designation and can be different for each site and could be different on either side of the project. In this case, twenty feet of minimum landscape buffer is required on the north and the south side along the property lines.

9.4. Stormwater Design Parameters

The most challenging part of designing and preparing a site engineering and construction plan is perhaps related to a stormwater management system. This discipline consumes most property and requires meeting or exceeding multiple agencies' rules and regulations. That is why it is crucial for the engineer and the designer to understand the requirements related to the property before diving deep into the design.

The following stormwater management design criteria are defined for this project as:

1) Water Quality Treatment (Dry Detention Pond)

This project was designed as a dry detention pond, since the pond bottom was set a minimum of two feet above the SHWE. The requirement was to provide online storage volume based on a half-inch rainfall over the project area for the water quality treatment.

2) Water Quantity Requirement (open basin)

The requirement is to meet or exceed the twenty-five-year, twenty-four-hour peak discharge since the project is within an open drainage basin. In other words, we need to ensure that storm runoff release from the site during post-development is equal or preferably blows the pre-development peak discharge for a twenty-five-year, twenty-four-hour storm event for our specific location.

Two (2) interconnected dry detention ponds will be constructed along the west property line to meet water quality and quantity requirements. A control structure will be built on the northwest corner in Pond 1 to control the discharge after the pond's water elevation reaches the required design elevation without altering the final discharge point of the system.

9.5. Required Approval/Permits

The following is a list of required permits for this project.

1) City permits from the planning and zoning, transportation (concurrency and access management), stormwater, utilities, and public works (ROW and traffic services), and fire departments.

2) Water management district for the stormwater management system,

3) Department of Health permit for the potable water testing and swimming pool and spa design and construction,

4) National Pollutant Discharge Elimination System (NPDES) notice of intent (NOI) construction permit,

5) Department of Environmental Protection Water and Wastewater,

6) Crime Prevention Through Environmental Design (CPTED) from the police department.

9.6. Final Site Layout

This project is about an acre smaller than Project 3 in Chapter 8, but the stormwater design concept will be the same with a caveat. This project is commercial, and many of its requirements are particular to a commercial property compared to a residential property. For example, in Project 3, we designed residential lots, and each single-family residence had its garage for parking. In this project, the parking area is public, and many more of them are required, including the accessible parking spaces and accessibility from those parking spaces to the building. Another example would be a commercial swimming pool and spa. Although, a SFR building may have a swimming pool and spa. Design, permitting, and construction of a public swimming pool and spa for a hotel have more restricted criteria to secure a permit through AHJ.

In the example of Project 3, we mentioned that the primary controlling discipline in the design is the stormwater management system and how it takes the most property. In this example, the stormwater design, number of parking spaces and the parking layout, driveways for traffic maneuverability on site, fire and solid waste truck access, swimming pool and spa location, and their pieces of equipment create more challenges. Remember that all these facilities must be designed and oriented to meet the minimum applicable governmental agencies' requirements. In addition, chain hotels must meet the minimum standards imposed by the franchise, which could be much more stringent. Regardless, all facilities must be safe and easily accessible to the public and the hotel customers. The long-term success of the hotel is directly related to their client's positive experiences. The ongoing success of such a facility depends on the client returning to this location repeatedly and doing word-of-mouth advertisements for the owner.

Let's understand the methodology presented in Chapter 8 and expand on it to design and prepare construction plans for any similar project. To include all computations and other related documentation here for this project is voluminous and can become a book in itself. Therefore, we don't go through all the calculations and construction plans and preparations or describe each construction plan sheet to save time and space. Those processes are the same as Project 3, as provided in Chapter 8. We mainly focus on providing some of the following pages to describe the existing site conditions and the final product's proposed layout.

9.7. Final Construction Plans Preparation

1) Sheet 1 of 17 - Cover Sheet

Reviewing thousands of projects, I noticed that not every engineer is keen on providing a cover sheet. Often, construction plans are submitted with lots of engineering data and other important information jammed in one or two sheets. It takes much longer to review these types of construction plans. A cover sheet with an index of sheets (even for two or three sheets of construction plans), project title, address, vicinity map, contact information such as the owner, engineer, architect, and utility companies' contact information should be an essential part of any construction plan set. The index of sheets (like a table of contents of a book) includes each sheet number and corresponding title.

A state map of where the project's location is shown on this sheet. Although, this is not necessary for small projects. A location or vicinity map with section, township, and range is more appropriate.

Provide the project title, owners, surveyor, engineer, and the contractor's contact information on this sheet. Often during the design and permitting process, the contractor is not known, and that is why we identified the contractor to be determined on this project. Utility companies' contact information can also be beneficial to show on this sheet.

Call Before You Dig is a vital utility locate phone number that may differ in each state and should be included at least on the cover sheet, but preferably on all construction plan sheets.

Sheets 1, 4, 5, 8, 16, and 17 of 17 at the end of this chapter represent the partial construction plan sheets for this project.

2) Sheet 4 of 17 - Site Plan Sheet

The site plan should include both the existing and the proposed information and site data to provide a realistic picture of the site conditions for the AHJ to review. It should also offer sufficient geometric dimensions for structures, envelopes, driveways, roadways, curbs, gutters, sidewalks, cul-de-sacs, landscape islands, and ponds for the surveyor and the contractor to perform the field's horizontal layout.

This sheet project layout includes the building envelope, parking layout, driveway accesses with drive isles, proposed ponds, and future development areas are shown with the exact dimensions. Project location, zoning information, project site data, building square footage for each floor, setbacks, parking information, flood zone information, signage, and FAR are tabulated under the building and site data.

3) Sheet 5 of 17 - Paving, Grading and Drainage Plan

This sheet may be the more challenging sheet in the entire set to compute correctly for an optimum design. It should have sufficient elevations to define the proposed site topography and contours. It should include structures envelopes, corner elevations, driveway accesses, drive aisles, curb and walkway, landscape islands, and ponds for the surveyor and the contractor to perform the field's horizontal layout. Existing public streets surround this site, which controls the matching elevations for the site. All other elevations between the property lines are determined by grading the site and how the proposed stormwater sheet flows to the proposed inlets and the designed pipes conveyed to the proposed ponds.

Two interconnected ponds are shown on this sheet. The proposed contours to include bottom and top elevations, design high water elevation, seasonal high groundwater elevation, soil boring locations, and the control structure dimensions and elevations are also indicated on this sheet.

Multiple elevations are proposed to ensure that the stormwater runoff is properly directed around the buildings and away from them. It is good to provide a table or spreadsheet for the proposed stormwater structures, such as S-1 and S-2, related to storm sewer pipes, elevations, and dimensions on the larger projects. The site parking lot is graded such that the stormwater runoff sheet flow toward the center of the driveways to the proposed inlets and via the new storm sewer system directed to the proposed ponds.

Proposed contours for the proposed ponds are shown on this sheet with the locations of the proposed cross-sections. Sheets 6 and 7 (not included) provide various sections A-A through F-F on the site to provide additional details and data throughout the site.

A storm sewer stub-out (capped) is shown on the southwest corner and another on the northwest corner of the site for the future improvements of storm sewer connections. This sheet also indicates the future restaurant location with a note "future improvements," which will be permitted separately. However, as mentioned earlier, if the total impervious area for the site stays below the assumed value of seventy percent, then there is no need to revise and resubmit the SWFWMD to modify the permit for revision.

Since the paving, grading, and drainage plan occupy the entire sheet, other related information and data are provided on other sheets, as indicated on the index of sheets. Soil profile at each pond and the actual soil properties, cross-sections, erosion, and sediment control during construction are presented on other sheets.

4) Sheet 8 of 17 - Utilities and Traffic Plan

The proposed waterline is connected to the existing six-inch water main located on the project's south side. An eight-inch by six-inch tapping sleeve and a gate valve are used to make the connection. The utility purveyor typically performs the connection to the mail line. This connection will provide a six-inch PVC waterline to the proposed three-story chain hotel. The six-inch service line is the minimum size required by the AHJ for commercial projects.

An existing six-inch gravity sewer line is located on the same street, perhaps unique to this location. Most AHJ does not allow smaller than an eight-inch gravity line in the public ROW. The existing six-inch sewer line was probably built long before more stringent criteria were enacted. Regardless, a manhole must be constructed at the connection point to connect to this mainline. The six-inch gravity service line will connect to the proposed sanitary manhole. A direct connection to the main gravity line will typically not be permitted by the AHJ.

This sheet will provide traffic signs, striping, etc., as related to permanent traffic control. All these devices must be per the *Manual of Uniform Traffic Control Devices (MUTCD)* and other AHJ standards and requirements.

Water, sanitary, and traffic related details and specifications are provided on other sheets indicated on the index of the sheets. Larger projects should have more elaborate water, sanitary sewer, and traffic signage plans and details for each discipline.

It is emphasized here that utility conflicts are essential to remember in utility design. Any permit that entails digging should include an 811 Call Before You Dig label in Florida, but this is especially important for utilities. This number may be different in other states or regions. It is 411 in the State of Washington. When looking at proposed utilities, it is crucial to consider both the horizontal and vertical position of all utilities along the property's frontage to ensure no proposed installations occupy the same space. If possible, a plan sheet should show this information for all utilities to make it readily apparent where utilities should be located relative to each other.

5) Sheet 16 of 17 - Pre-development Drainage Basin Plan

The primary purpose of this and the next sheet is to provide pre-development (existing site conditions) drainage information to facilitate drainage computation and report. They are not part of construction plan sheets and do not need to be included in construction plans offered to contractors for bidding, except for information only. However, it is included in the drainage report submitted to permitting agencies reviewing and approving stormwater management design.

Let us review some of the information provided on this sheet.

The pre-development drainage area is subdivided into two smaller subbasin areas, 1 and 2. Subbasin area 1 covers more than eighty percent of the site. Stormwater runoff sheet flows from the contour line at an approximate elevation of 154 feet to the northwest corner of the property. The rest of the project area sheet flows to the east and eventually reaches the property's southeast corner. The basin and subbasin boundaries, drainage areas in acreage, and hydraulic length are shown on this sheet. In the drainage report (not included here), a concentration-time of 19.54 minutes was computed for the larger subbasin area 1. For the smaller subbasin area 2, the concentration-time is assumed to be ten minutes.

6)   Sheet 17 of 17 - Post-development Drainage Basin Plan

Many spot elevations must be proposed to properly direct stormwater runoff from the project area, around the proposed facilities to the proposed inlets, to storm sewer pipes, and eventually to proposed ponds. Therefore, based on the proposed elevations, the post-development basin boundary is subdivided into eleven smaller subbasin areas numbered 1 to 11. We assumed a concentration time of fifteen minutes for the larger areas 1, 2, 3, 4, 5, and 6. A ten-minute concentration time for the rest of the sub-areas 7, 8, 9, 10, and 11 was assumed. To be more conservative, one can assume a minimum of ten minutes concentration for each area.

A conservative assumption means a lower concentration-time, which means it takes a shorter time for a raindrop to travel from point A to point B. The faster the travel time means less infiltration if traveling through the porous surfaces such as dirt and grass area. If it travels over impervious surface areas, such as asphalt or concrete pavement, it means less evaporation. The result in either scenario is the higher volume accumulation and peak discharge.

The following pages represent partial construction plans sheet 1, 4, 5, 8, 16 and 17 of 17.

SITE CONSTRUCTION PLANS

FOR

PROPOSED FRANCHISE HOTEL

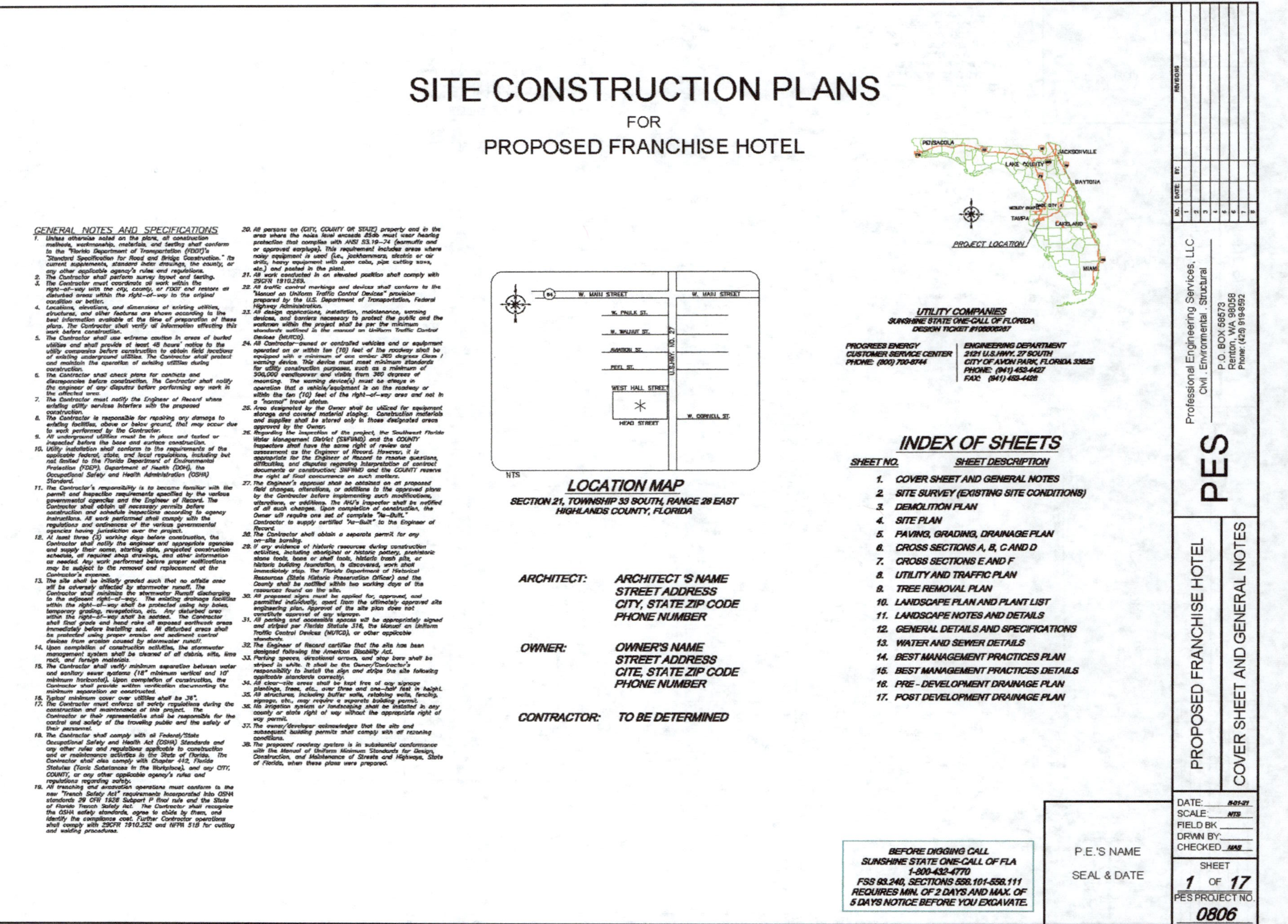

GENERAL NOTES AND SPECIFICATIONS

1. Unless otherwise noted on the plans, all construction methods, workmanship, materials, and testing shall conform to the "Florida Department of Transportation (FDOT)'s "Standard Specification for Road and Bridge Construction." Its current supplements, standard index drawings, the county, or any other applicable agency's rules and regulations.
2. The Contractor shall perform survey layout and testing.
3. The Contractor must coordinate all work within the right-of-way with the city, county, or FDOT and restore all disturbed areas within the right-of-way to the original condition or better.
4. Locations, elevations, and dimensions of existing utilities, structures, and other features are shown according to the best information available at the time of preparation of these plans. The Contractor shall verify all information effecting this work before construction.
5. The Contractor shall use extreme caution in areas of buried utilities and shall provide at least 48 hours' notice to the utility companies before construction to obtain field locations of existing underground utilities. The Contractor shall protect and maintain the operation of existing utilities during construction.
6. The Contractor shall check plans for conflicts and discrepancies before construction. The Contractor shall notify the engineer of any disputes before performing any work in the affected area.
7. The Contractor must notify the Engineer of Record where existing utility services interfere with the proposed construction.
8. The Contractor is responsible for repairing any damage to existing facilities, above or below ground, that may occur due to work performed by the Contractor.
9. All underground utilities must be in place and tested or inspected before the base and surface construction.
10. Utility installation shall conform to the requirements of the applicable federal, state, and local regulations, including but not limited to the Florida Department of Environmental Protection (FDEP), Department of Health (DOH), the Occupational Safety and Health Administration (OSHA) Standard.
11. The Contractor's responsibility is to become familiar with the permit and inspection requirements specified by the various governmental agencies and the Engineer of Record. The Contractor shall obtain all necessary permits before construction and schedule inspections according to agency instructions. All work performed shall comply with the regulations and ordinances of the various governmental agencies having jurisdiction over the project.
12. At least three (3) working days before construction, the Contractor shall notify the engineer and appropriate agencies and supply their name, starting date, projected construction schedule, all required shop drawings, and other information as needed. Any work performed before proper notifications may be subject to the removal and replacement at the Contractor's expense.
13. The site shall be initially graded such that no offsite area will be adversely affected by stormwater runoff. The Contractor shall minimize the stormwater Runoff discharging to the adjacent right-of-way. The existing drainage facilities within the right-of-way shall be protected using hay bales, temporary grading, revegetation, etc. Any disturbed area within the right-of-way shall be sodded. The Contractor shall final grade and hand rake all exposed earthwork areas immediately before installing sod. All disturbed areas shall be protected using proper erosion and sediment control devices from erosion caused by stormwater runoff.
14. Upon completion of construction activities, the stormwater management system shall be cleaned of all debris, silts, lime rock, and foreign materials.
15. The Contractor shall verify minimum separation between water and sanitary sewer systems (18" minimum vertical and 10' minimum horizontal). Upon completion of construction, the Contractor shall provide written verification documenting the minimum separation as constructed.
16. Typical minimum cover over utilities shall be 36".
17. The Contractor must enforce all safety regulations during the construction and maintenance of this project. The Contractor or their representative shall be responsible for the control and safety of the traveling public and the safety of their personnel.
18. The Contractor shall comply with all Federal/State Occupational Safety and Health Act (OSHA) Standards and any other rules and regulations applicable to construction and or maintenance activities in the State of Florida. The Contractor shall also comply with Chapter 442, Florida Statutes (Toxic Substances in the Workplace), and any CITY, COUNTY, or any other applicable agency's rules and regulations regarding safety.
19. All trenching and excavation operations must conform to the new "Trench Safety Act" requirements incorporated into OSHA standards 29 CFR 1926 Subpart P final rule and the State of Florida Trench Safety Act. The Contractor shall recognize the OSHA safety standards, agree to abide by them, and identify the compliance cost. Further Contractor operations shall comply with 29CFR 1910.252 and NFPA 51B for cutting and welding procedures.
20. All persons on (CITY, COUNTY OR STATE) property and in the area where the noise level exceeds 85db must wear hearing protection that complies with ANSI S3.19-74 (earmuffs and or approved earplugs). This requirement includes areas where noisy equipment is used (i.e., jackhammers, electric or air drills, heavy equipment with open cabs, pipe cutting saws, etc.) and posted in the plant.
21. All work conducted in an elevated position shall comply with 29CFR 1910.269.
22. All traffic control markings and devices shall conform to the "Manual on Uniform Traffic Control Devices" provision prepared by the U.S. Department of Transportation, Federal Highway Administration.
23. All design applications, installation, maintenance, warning devices, and barriers necessary to protect the public and the workmen within the project shall be per the minimum standards outlined in the manual on Uniform Traffic Control Devices (MUTCD).
24. All Contractor-owned or controlled vehicles and or equipment operated on or within ten (10) feet of the roadway shall be equipped with a minimum of one amber 360 degrees Class I warning device. This device must meet minimum standards for utility construction purposes, such as a minimum of 500,000 candlepower and visible from 360 degrees of mounting. The warning device(s) must be always in operation that a vehicle/equipment is on the roadway or within the ten (10) feet of the right-of-way area and not in a "normal" travel status.
25. Area designated by the Owner shall be utilized for equipment storage and covered material staging. Construction materials and supplies shall be stored only in those designated areas approved by the Owner.
26. Regarding the inspection of the project, the Southwest Florida Water Management District (SWFWMD) and the COUNTY Inspectors shall have the same right of review and assessment as the Engineer of Record. However, it is appropriate for the Engineer of Record to resolve questions, difficulties, and disputes regarding interpretation of contract documents or construction; SWFWMD and the COUNTY reserve the right of final concurrence on such matters.
27. The Engineer's approval shall be obtained on all proposed field changes, alterations, or additions to the approved plans by the Contractor before implementing such modifications, alterations, or additions. The AHJ's inspector shall be notified of all such changes. Upon completion of construction, the Owner will require one set of complete "As-Built." Contractor to supply certified "As-Built" to the Engineer of Record.
28. The Contractor shall obtain a separate permit for any on-site burning.
29. If any evidence of historic resources during construction activities, including aboriginal or historic pottery, prehistoric stone tools, bone or shell tools, historic trash pits, or historic building foundation, is discovered, work shall immediately stop. The Florida Department of Historical Resources (State Historic Preservation Officer) and the County shall be notified within two working days of the resources found on the site.
30. All proposed signs must be applied for, approved, and permitted individually, apart from the ultimately approved site engineering plan. Approval of the site plan does not constitute approval of any signage.
31. All parking and accessible spaces will be appropriately signed and striped per Florida Statute 316, the Manual on Uniform Traffic Control Devices (MUTCD), or other applicable standards.
32. The Engineer of Record certifies that the site has been designed following the American Disability Act.
33. Parking spaces, directional arrows, and stop bars shall be striped in white. It shall be the Owner/Contractor's responsibility to install the sign and stripe the site following applicable standards correctly.
34. All clear-site areas shall be kept free of any signage plantings, trees, etc., over three and one-half feet in height.
35. All structures, including buffer walls, retaining walls, fencing, signage, etc., may require a separate building permit.
36. No irrigation system or landscaping shall be installed in any county or state right of way without the appropriate right of way permit.
37. The owner/developer acknowledges that the site and subsequent building permits shall comply with all rezoning conditions.
38. The proposed roadway system is in substantial conformance with the Manual of Uniform Minimum Standards for Design, Construction, and Maintenance of Streets and Highways, State of Florida, when these plans were prepared.

ARCHITECT: ARCHITECT'S NAME
STREET ADDRESS
CITY, STATE ZIP CODE
PHONE NUMBER

OWNER: OWNER'S NAME
STREET ADDRESS
CITE, STATE ZIP CODE
PHONE NUMBER

CONTRACTOR: TO BE DETERMINED

INDEX OF SHEETS

SHEET NO.	SHEET DESCRIPTION
1.	COVER SHEET AND GENERAL NOTES
2.	SITE SURVEY (EXISTING SITE CONDITIONS)
3.	DEMOLITION PLAN
4.	SITE PLAN
5.	PAVING, GRADING, DRAINAGE PLAN
6.	CROSS SECTIONS A, B, C AND D
7.	CROSS SECTIONS E AND F
8.	UTILITY AND TRAFFIC PLAN
9.	TREE REMOVAL PLAN
10.	LANDSCAPE PLAN AND PLANT LIST
11.	LANDSCAPE NOTES AND DETAILS
12.	GENERAL DETAILS AND SPECIFICATIONS
13.	WATER AND SEWER DETAILS
14.	BEST MANAGEMENT PRACTICES PLAN
15.	BEST MANAGEMENT PRACTICES DETAILS
16.	PRE - DEVELOPMENT DRAINAGE PLAN
17.	POST DEVELOPMENT DRAINAGE PLAN

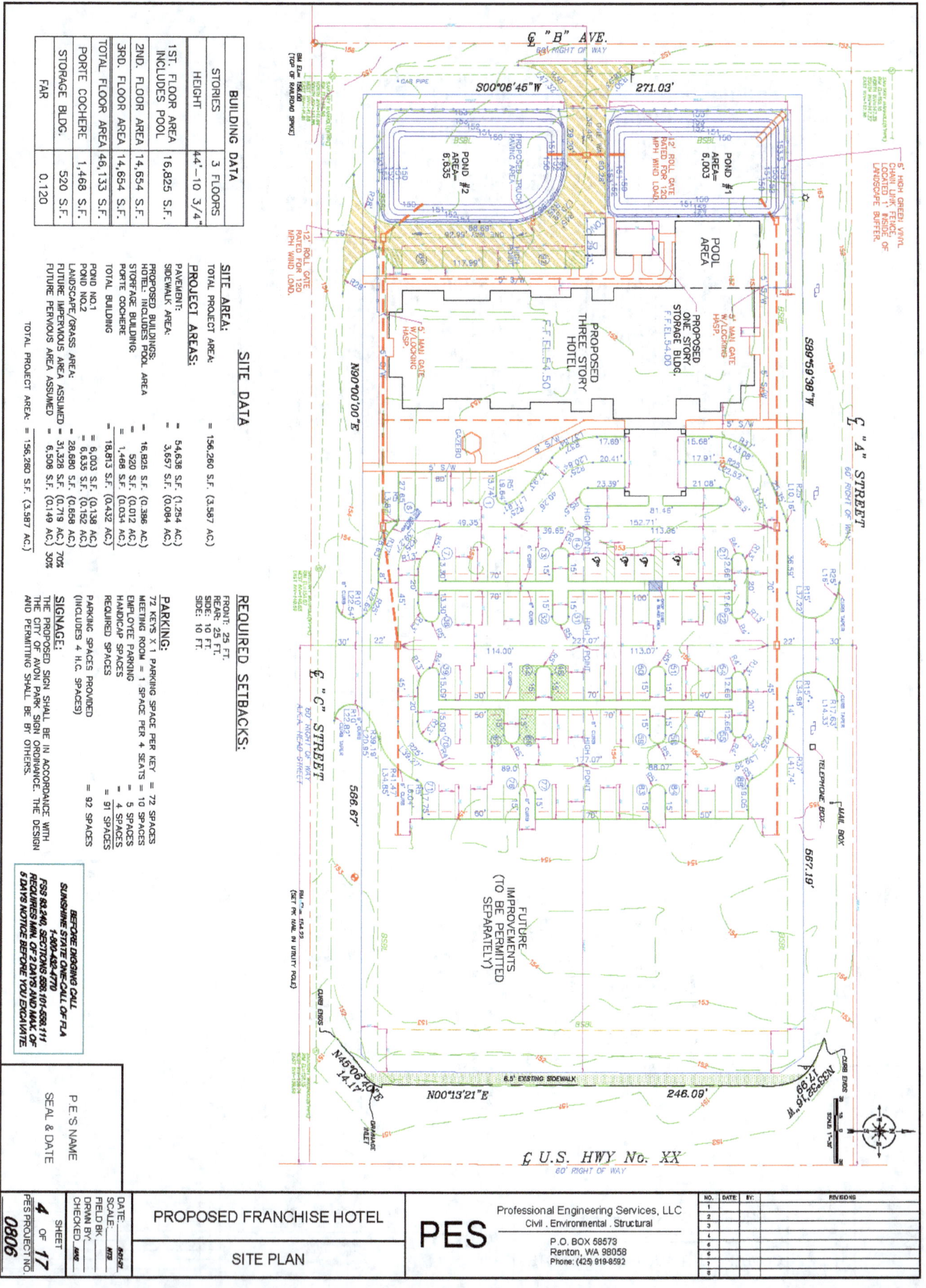
BUILDING DATA
STORIES — 3 FLOORS
HEIGHT — 44'-10 3/4"
1ST. FLOOR AREA INCLUDES POOL — 16,825 S.F.
2ND. FLOOR AREA — 14,654 S.F.
3RD. FLOOR AREA — 14,654 S.F.
TOTAL FLOOR AREA — 46,133 S.F.
PORTE COCHERE — 1,468 S.F.
STORAGE BLDG. — 520 S.F.
FAR — 0.120

SITE DATA

SITE AREA:
TOTAL PROJECT AREA: = 156,260 S.F. (3.587 AC.)

PROJECT AREAS:
PAVEMENT: = 54,636 S.F. (1.254 AC.)
SIDEWALK AREA: = 3,657 S.F. (0.084 AC.)

PROPOSED BUILDINGS:
HOTEL: INCLUDES POOL AREA = 16,825 S.F. (0.386 AC.)
STORAGE BUILDING: = 520 S.F. (0.012 AC.)
PORTE COCHERE = 1,468 S.F. (0.034 AC.)
TOTAL BUILDING = 18,813 S.F. (0.432 AC.)

POND NO.1 = 6,003 S.F. (0.138 AC.)
POND NO.2 = 6,635 S.F. (0.152 AC.)
LANDSCAPE/GRASS AREA: = 28,680 S.F. (0.658 AC.)
FUTURE IMPERVIOUS AREA ASSUMED = 31,328 S.F. (0.719 AC.) 70%
FUTURE PERVIOUS AREA ASSUMED = 6,508 S.F. (0.149 AC.) 30%

TOTAL PROJECT AREA: = 156,260 S.F. (3.587 AC.)

REQUIRED SETBACKS:
FRONT: 25 FT.
REAR: 25 FT.
SIDE: 10 FT.
SIDE: 10 FT.

PARKING:
72 KEYS X 1 PARKING SPACE PER KEY = 72 SPACES
MEETING ROOM = 1 SPACE PER 4 SEATS = 10 SPACES
EMPLOYEE PARKING = 5 SPACES
HANDICAP SPACES = 4 SPACES
REQUIRED SPACES = 91 SPACES
PARKING SPACES PROVIDED = 92 SPACES
(INCLUDES 4 H.C. SPACES)

SIGNAGE:
THE PROPOSED SIGN SHALL BE IN ACCORDANCE WITH
THE CITY OF AVON PARK SIGN ORDINANCE. THE DESIGN
AND PERMITTING SHALL BE BY OTHERS.

BEFORE DIGGING CALL
SUNSHINE STATE ONE-CALL OF FLA
1-800-432-4770
FSS 83.240, SECTIONS 556.101-556.111
REQUIRES MIN. OF 2 DAYS AND MAX. OF
5 DAYS NOTICE BEFORE YOU EXCAVATE.

PROPOSED FRANCHISE HOTEL
SITE PLAN
PES
Professional Engineering Services, LLC
Civil . Environmental . Structural
P.O. BOX 58573
Renton, WA 98058
Phone: (425) 919-8592

DATE:
SCALE:
FIELD BK:
DRWN BY:
CHECKED
SHEET 4 OF 17
PES PROJECT NO. 0806

P.E.'S NAME
SEAL & DATE

POOL AREA
PROPOSED THREE STORY HOTEL
F.F.EL.54.50
PROPOSED ONE STORY STORAGE BLDG. F.F.EL.54.00
POND #1 AREA= 6,003
POND #2 AREA= 6,635
PROPOSED TRUCK PAVING AREA
GAZEBO
FUTURE IMPROVEMENTS (TO BE PERMITTED SEPARATELY)
5' MAU GATE W/LOCKING HASP.
12' ROLL GATE RATED FOR 120 MPH WIND LOAD.
6' HIGH GREEN VINYL CHAIN LINK FENCE, LOCATED INSIDE OF LANDSCAPE BUFFER.

C "B" AVE. 60' RIGHT OF WAY
C "A" STREET 60' RIGHT OF WAY
C "C" STREET 60' RIGHT OF WAY AKA HEAD STREET
C U.S. HWY No. XX 60' RIGHT OF WAY
6.5' EXISTING SIDEWALK
DRAINAGE INLET
TELEPHONE BOX
MAIL BOX

S00°06'45"W 271.03'
N90°00'00"E
S89°59'38"W
N00°13'21"E 246.09'
N45°06'40"E 14.17'
566.67'
567.19'

SCALE 1"=30'

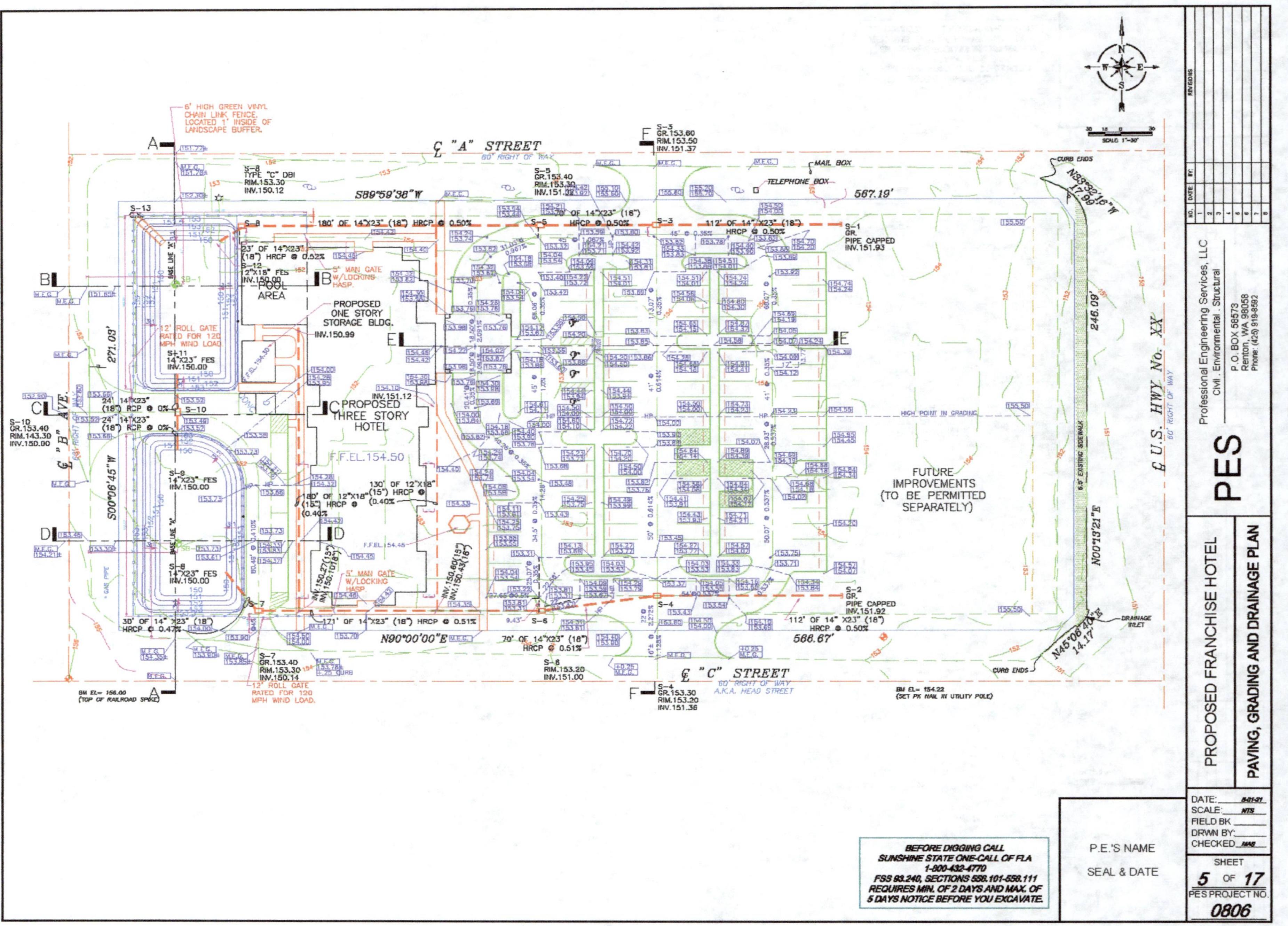

SITE DESIGN & ENGINEERING
PROPOSED FRANCHISE HOTEL
PAVING, GRADING AND DRAINAGE PLAN
PES
Professional Engineering Services, LLC
Civil . Environmental . Structural
P.O. BOX 58573
Renton, WA 98058
Phone (425) 919-8592
DATE:
SCALE: NTS
FIELD BK
DRWN BY:
CHECKED
P.E.'S NAME
SEAL & DATE
SHEET 5 OF 17
PES PROJECT NO. 0806
BEFORE DIGGING CALL
SUNSHINE STATE ONE-CALL OF FLA
1-800-432-4770
FSS 93.240, SECTIONS 556.101-556.111
REQUIRES MIN. OF 2 DAYS AND MAX. OF
5 DAYS NOTICE BEFORE YOU EXCAVATE.
REVISIONS
NO. DATE: BY:
6' HIGH GREEN VINYL CHAIN LINK FENCE, LOCATED 1' INSIDE OF LANDSCAPE BUFFER.
C "A" STREET
60' RIGHT OF WAY
C "C" STREET
60' RIGHT OF WAY
A.K.A. HEAD STREET
C "B" AVE.
60' RIGHT OF WAY
E U.S. HWY No. XX
60' RIGHT OF WAY
6.5' EXISTING SIDEWALK
PROPOSED ONE STORY STORAGE BLDG.
INV.150.99
PROPOSED THREE STORY HOTEL
F.F. EL. 154.50
F.F. EL. 154.45
POOL AREA
FUTURE IMPROVEMENTS (TO BE PERMITTED SEPARATELY)
HIGH POINT IN GRADING
MAIL BOX
TELEPHONE BOX
DRAINAGE INLET
CURB ENDS
567.19'
566.67'
246.09'
271.03'
S89°59'38"W
N90°00'00"E
S00°06'45"W
N00°13'21"E
N45°06'40"E 14.17'
N33°32'16"W 17.99'
SCALE 1"=30'
S-3 GR.153.60 RIM.153.50 INV.151.37
S-5 GR.153.40 RIM.153.30 INV.151.92
S-8 TYPE "C" DBI RIM.153.30 INV.150.12
S-12 12"X18" FES INV.150.00
S-11 14"X23" FES INV.150.00
S-10 GR.153.40 RIM.143.30 INV.150.00
S-9 14"X23" FES INV.150.00
S-8 14"X23" FES INV.150.00
S-7 GR.153.40 RIM.153.30 INV.150.14
S-6 RIM.153.20 INV.151.00
S-4 GR.153.30 RIM.153.20 INV.151.36
S-2 GR. PIPE CAPPED INV.151.92
S-1 GR. PIPE CAPPED INV.151.93
S-13
180' OF 14"X23" (18") HRCP @ 0.50%
112' OF 14"X23" (18") HRCP @ 0.50%
23' OF 14"X23" (18") HRCP @ 0.52%
24' 14"X23" (18") RCP @ 0%
24' 14"X23" (18") RCP @ 0%
130' OF 12"X18" (15") HRCP @ 0.40%
18' OF 12"X18" (15") HRCP @ 0.40%
171' OF 14"X23" (18") HRCP @ 0.51%
70' OF 14"X23" (18") HRCP @ 0.51%
30' OF 14"X23" (18") HRCP @ 0.47%
112' OF 14"X23" (18") HRCP @ 0.50%
INV.151.12
INV.150.27(15")
INV.150.10(18")
INV.150.80(15")
INV.150.43(18")
5' MAN GATE W/LOCKING HASP.
5' MAN GATE W/LOCKING HASP.
12' ROLL GATE RATED FOR 120 MPH WIND LOAD.
12' ROLL GATE RATED FOR 120 MPH WIND LOAD.
BASE LINE "X"
BASE LINE "X"
BM EL.= 156.00 (TOP OF RAILROAD SPIKE)
BM EL.= 154.22 (SET PK NAIL IN UTILITY POLE)
CURB ENDS

ALI SHASTI, P.E.

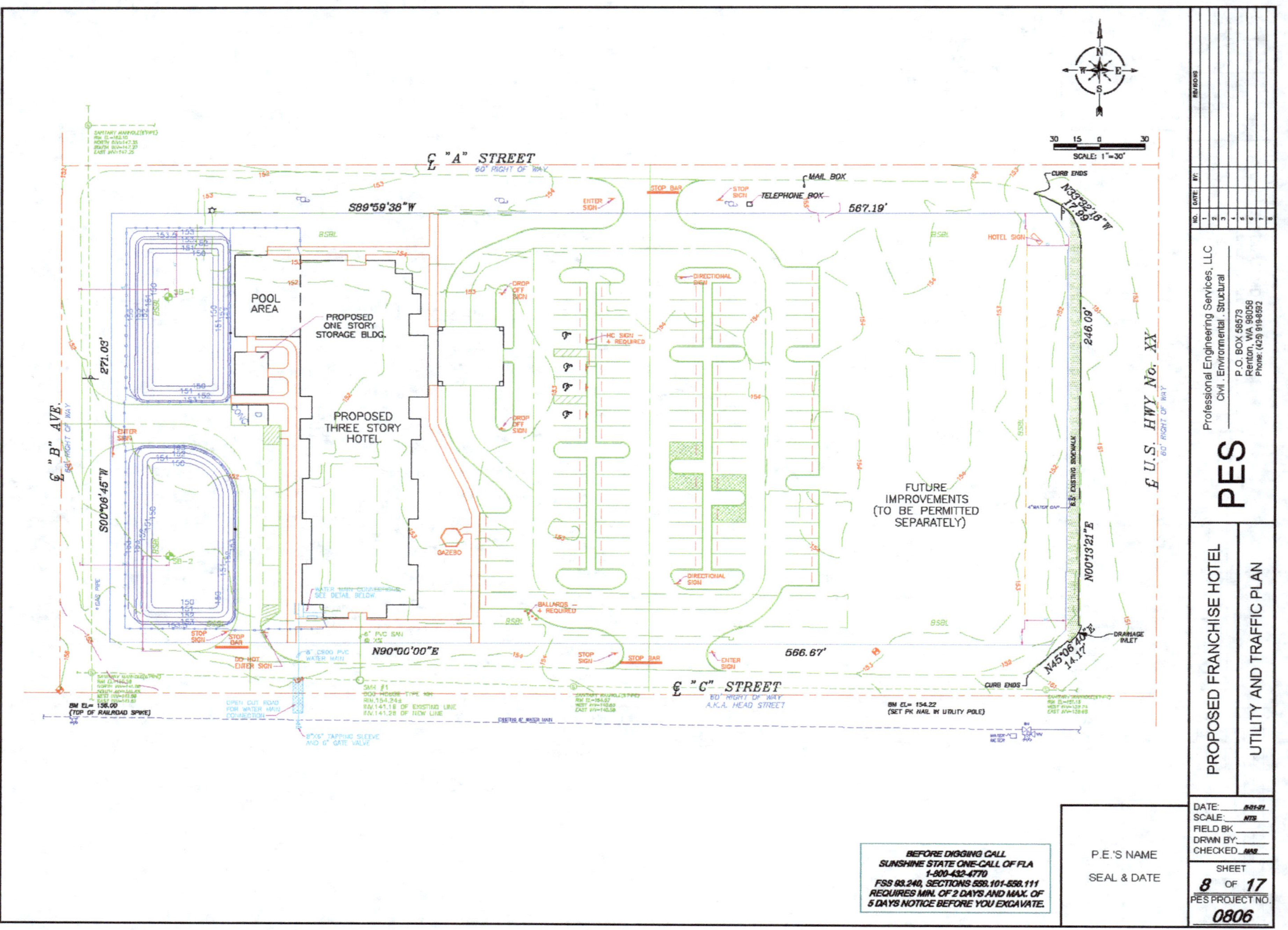

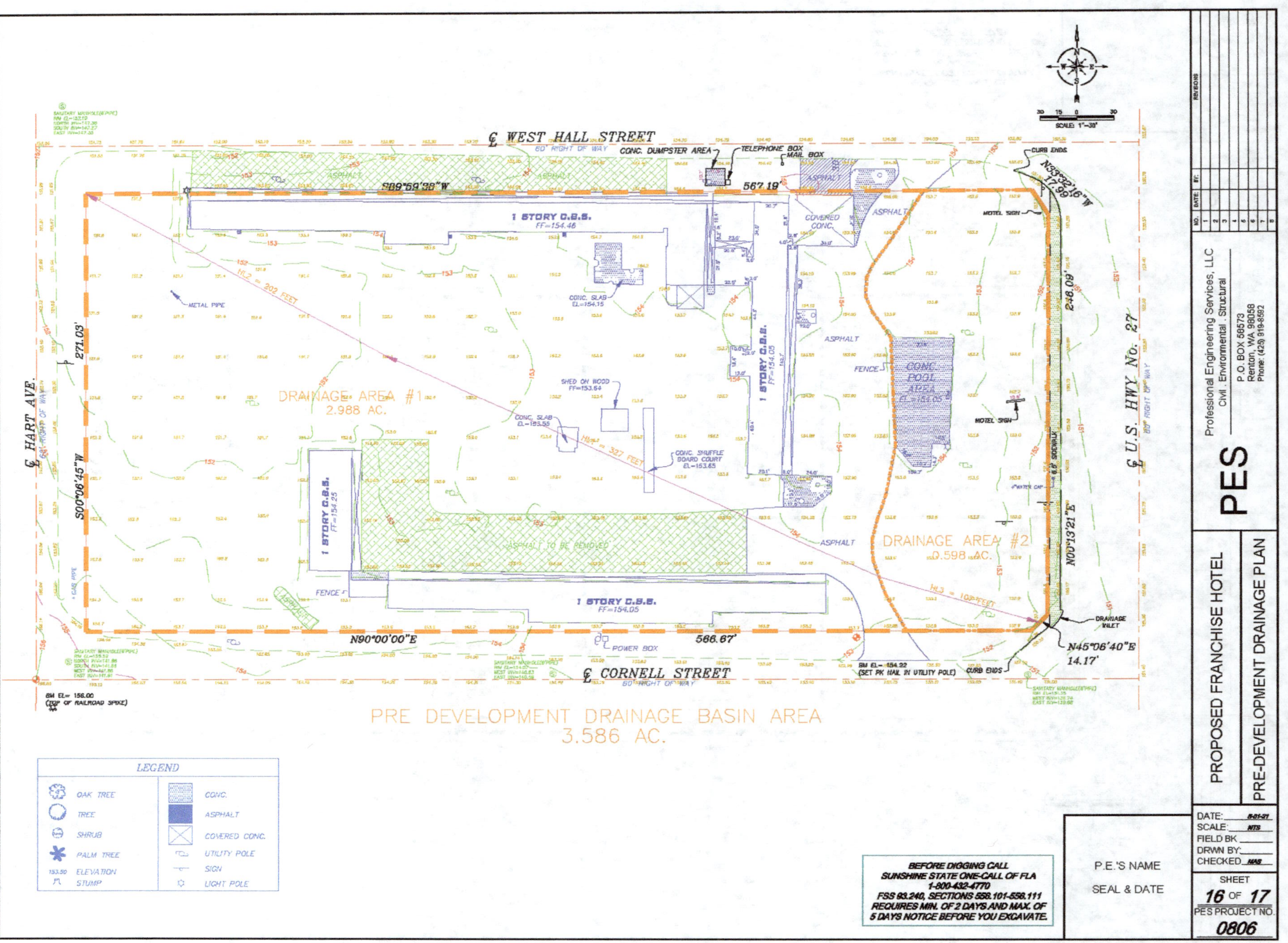

SITE DESIGN & ENGINEERING
PROPOSED FRANCHISE HOTEL
PRE-DEVELOPMENT DRAINAGE PLAN
PES
Professional Engineering Services, LLC
Civil. Environmental. Structural
P.O. BOX 58573
Renton, WA 98058
Phone: (425) 919-8592
DATE: 8-01-21
SCALE: NTS
FIELD BK
DRWN BY:
CHECKED: MAS
SHEET 16 OF 17
PES PROJECT NO. 0806
P.E.'S NAME
SEAL & DATE
BEFORE DIGGING CALL
SUNSHINE STATE ONE-CALL OF FLA
1-800-432-4770
FSS 93.240, SECTIONS 556.101-556.111
REQUIRES MIN. OF 2 DAYS AND MAX. OF
5 DAYS NOTICE BEFORE YOU EXCAVATE.
SCALE 1"=30'
WEST HALL STREET
60' RIGHT OF WAY
CONC. DUMPSTER AREA
TELEPHONE BOX
MAIL BOX
CURB ENDS.
567.19
S89°59'38"W
ASPHALT
COVERED CONC.
ASPHALT
MOTEL SIGN
N39°32'16"W 17.99
1 STORY C.B.S. FF=154.46
CONC. SLAB EL=154.15
1 STORY C.B.S. FF=154.05
FENCE
CONC. POOL AREA EL=154.05
MOTEL SIGN
METAL PIPE
HL2 = 209 FEET
DRAINAGE AREA #1
2.988 AC.
SHED ON WOOD FF=153.64
CONC. SLAB EL=153.59
HL3 = 327 FEET
CONC. SHUFFLE BOARD COURT EL=153.65
ASPHALT
DRAINAGE AREA #2
0.598 AC.
HART AVE.
68' RIGHT OF WAY
271.03'
S00°06'45"W
1 STORY C.B.S. FF=154.25
ASPHALT TO BE REMOVED
FENCE
1 STORY C.B.S. FF=154.05
N90°00'00"E
566.67'
POWER BOX
CORNELL STREET
60' RIGHT OF WAY
BM EL=154.22 (SET PK NAIL IN UTILITY POLE)
CURB ENDS.
N45°06'40"E 14.17'
DRAINAGE INLET
HL3 = 109 FEET
U.S. HWY No. 27
80' RIGHT OF WAY
246.09'
N00°13'21"E
6.5' SIDEWALK
BM EL=156.00 (TOP OF RAILROAD SPIKE)
PRE DEVELOPMENT DRAINAGE BASIN AREA
3.586 AC.
LEGEND
OAK TREE
TREE
SHRUB
PALM TREE
153.50 ELEVATION
STUMP
CONC.
ASPHALT
COVERED CONC.
UTILITY POLE
SIGN
LIGHT POLE
REVISIONS
NO. DATE BY
183

SCALE: 1"=30'
C "A" STREET
60' RIGHT OF WAY
MAIL BOX
TELEPHONE BOX
CURB ENDS
N89°32'16"W
17.99'
S89°59'58"W
587.19'
DRAINAGE AREA #11
0.207 AC.
POOL AREA
DRAINAGE AREA #8
0.161 AC.
PROPOSED ONE STORY STORAGE BLDG.
DRAINAGE AREA #5
0.332 AC.
DRAINAGE AREA #3
0.487 AC.
DRAINAGE AREA #1
0.498 AC.
TOP 0.138 AC.
BOT 0.070 AC.
DRAINAGE AREA #10
0.055 AC.
PROPOSED THREE STORY HOTEL
9.5' EXISTING SIDEWALK
248.09'
N00°13'21"E
FUTURE IMPROVEMENTS
(TO BE PERMITTED SEPARATELY)
DRAINAGE AREA #4
0.396 AC.
DRAINAGE AREA #2
0.401 AC.
C "B" AVE.
66' RIGHT OF WAY
DRAINAGE AREA #7
0.298 AC.
DRAINAGE AREA #9
0.217 AC.
TOP 0.152 AC.
BOT 0.082 AC.
DRAINAGE AREA #6
0.536 AC.
S00°06'45"W
521.03'
N90°00'00"E
566.67'
N45°06'40"E
14.17'
DRAINAGE INLET
CURB ENDS
C "C" STREET
60' RIGHT OF WAY
A.K.A. HEAD STREET
BM EL= 156.00
(TOP OF RAILROAD SPIKE)
BM EL= 154.22
(SET PK NAIL IN UTILITY POLE)
E U.S. HWY No. XX
60' RIGHT OF WAY
POST DEVELOPMENT DRAINAGE AREA = 3.586 AC.
Professional Engineering Services, LLC
Civil . Environmental . Structural
PES
P.O. BOX 68673
Renton, WA 98058
Phone: (425) 919-8592
PROPOSED FRANCHISE HOTEL
POST DEVELOPMENT DRAINAGE PLAN
REVISIONS
NO. DATE BY:
DATE: 5-21-21
SCALE: NTS
FIELD BK.
DRWN BY:
CHECKED MAS
SHEET
17 OF 17
PES PROJECT NO.
0806
P.E.'S NAME
SEAL & DATE
BEFORE DIGGING CALL
SUNSHINE STATE ONE-CALL OF FLA
1-800-432-4770
FSS 93.240, SECTIONS 556.101-556.111
REQUIRES MIN. OF 2 DAYS AND MAX. OF
5 DAYS NOTICE BEFORE YOU EXCAVATE.

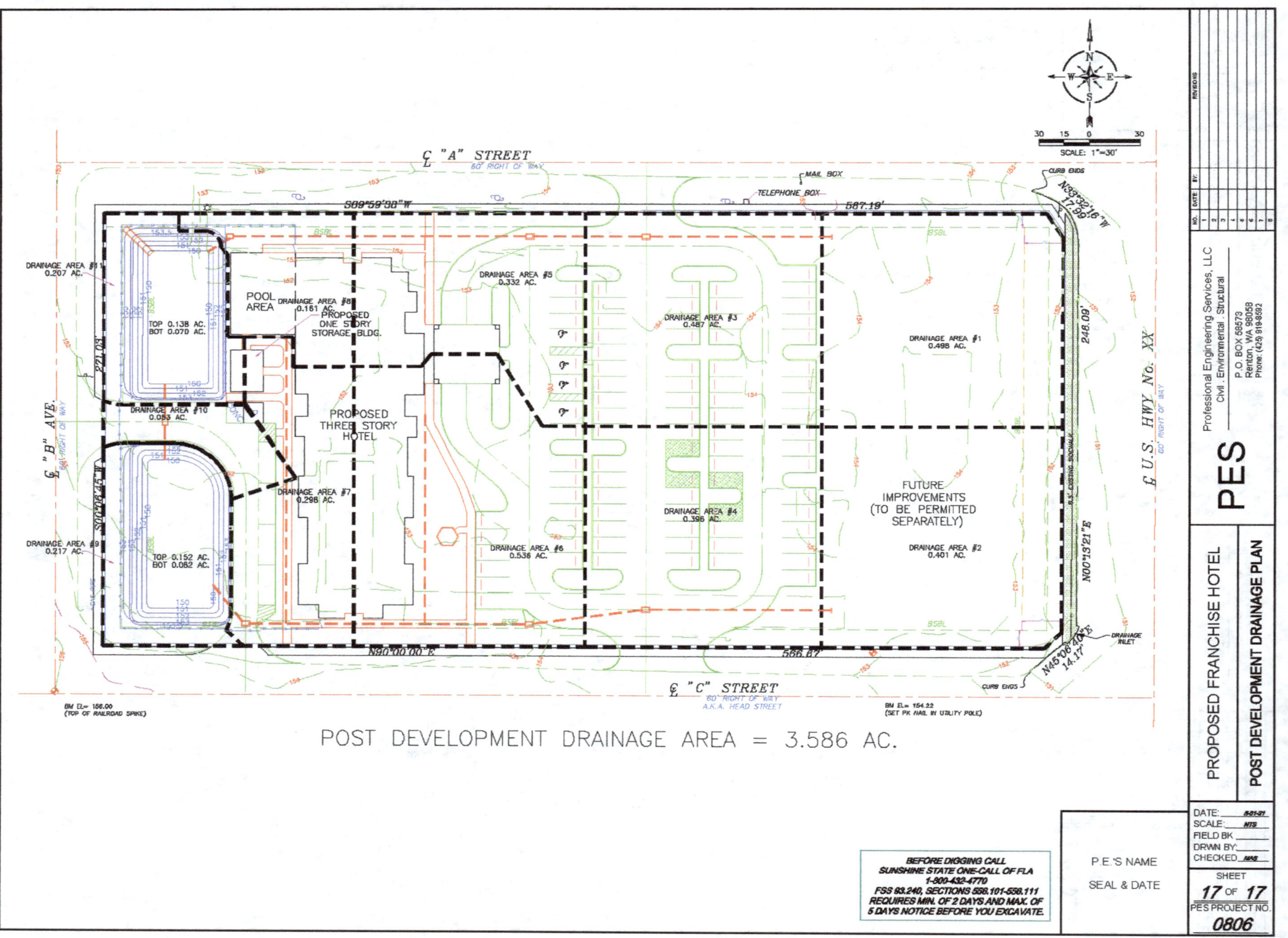

184

CHAPTER 10: AGENCIES SUBMITTAL, APPROVAL AND PERMITTING

10.1. Introduction

This chapter is perhaps the most critical chapter of this book! It is essential because regardless of one's effort to design and prepare a well-thought-out construction plan, no permits mean no construction or implementation of the developed project. To start and complete construction activities, agency permitting is one of the most critical steps. Unfortunately, there are no uniform standards related to various governmental agencies permitting. Many agencies have similar rules and regulations, mainly obtained from the same source or each other. Regardless, there are enough differences in laws and regulations to make permitting from each agency challenging.

This chapter's primary purpose is to guide the applicant from an early stage of the pre-submittal meeting to obtaining construction approval and securing permits, implementation, and submittal of the final as built or record drawing.

10.2. Permitting Process

The following page provides a flowchart with a general summary of the step by step permitting process. It is illustrative only. The following pages give a more descriptive and detailed explanation of each step. These steps may be reduced or more elaborate depending on the project's size and complexity and the permitting organization's size and sophistication. Typically, the larger and more established the organization, the rules and regulations become more extensive and restrictive. Although, the same organizations may provide more clear and concise instructions to their customers to streamline and help them navigate the system. Readers are encouraged to contact their respective agencies to obtain more specific nuances and up-to-date information about their projects.

PRE-APP
- Step 1: Meet with the AHJ to discuss preliminary/concept plans submitted by the applicant. Questions and answers.

INTAKE
- Step 2: Formal submittal of the construction plans and associated documents to the AHJ. It may have to be in person.

REVIEW
- Step 3: Plans and documents are routed to the reviewers for review and comments. Review time varies among agencies.

APPROVAL
- Step 4: The AJH approves plans and documents. This step may take up to a few iterations and possible follow-up meetings.

CONSTRUCTION
- Step 5: Permits are issued, and the construction starts. This step requires multiple meetings and coordination with the site inspector (s).

FINAL
- Step 6: The inspector (s) approve that the final site construction layout is built substantially per the approved construction plans and related documents.

Note: Many agencies do not issue a building permit until site engineering, ROW construction plans, and associated improvements are approved and ready to be given. Some agencies don't accept building plans until a copy of the approved site engineering and ROW plans are accompanied and submitted with the architectural building plans. Therefore, it behooves the applicant to prepare the site construction and ROW plans to Step 4 in the above flowchart before spending considerable time and resources preparing the final building architectural plan.

10.3. Pre-submittal Meeting (Step 1)

Pre-submittal or pre-application (aka, development review committee/DRC) meeting is an excellent way for the applicant, owner, engineer, or architect to obtain the basic information necessary to design a site or building. Some agencies may require a pre-submittal meeting for larger projects such as subdivisions, multi-residential and commercial projects and encourage smaller projects such as single-family residential (SFR). For an SFR, the applicant may skip the pre-submittal meeting and submit the application with construction drawings for review and approval. Regardless of the applicable agencies' requirements, it is time well spent to start a project by meeting reviewers and picking their brains before beginning a design.

Most agencies require the applicant to submit a preliminary plan; some may require completing a pre-application form and sometimes require a fee to schedule a meeting. Depending on the agency's policies and workload, a meeting may be scheduled within one or two days up to three to four weeks. To schedule a pre-submittal meeting, the applicant needs to call or e-mail the permit center and schedule an appointment.

The purpose of a pre-application meeting is to provide an opportunity for every department or division of the agency to review a preliminary plan proposed by an applicant. This way, each reviewer brings up potential issues and problems that could be detrimental to the project's success before the applicant or the owner spends tremendous time and financial resources on their project.

Agencies with a mandatory pre-submittal requirement usually ask that the applicant submit a preliminary plan and a completed pre-submittal application that includes basic information for the property. The permit center will distribute construction plans and applications to all reviewers for a concept review. Each reviewer brings comments and a list of project requirements to the meeting for discussion. It is essential that those who attend this meeting be prepared and familiar with the project and list questions and concerns concerning the project. Also, ensure each reviewer's business card or a list of attendees' contact information, especially phone numbers and e-mail addresses. Here is a short list of common questions that one should ask in case not already covered by one of the reviewers.

1. What are the fees (i.e., application, review, inspections)
2. Process (i.e., first review comments, resubmittal, revisions)
3. Existing utility information (i.e., water, wastewater, electric)
4. Required setback: front, back, and sides
5. Required landscape buffers: front, back, and sides

6. Are there any frontage improvements required? Details? How about the ROW dedication?
7. Is fee-in-lieu accepted for the frontage improvements? If yes, what is the process and timing for the payment?
8. The stormwater management criteria and if there is a fee-in-lieu accepted for this. If yes, what is the process and timing for the payment?
9. Other questions specific to the project.

10.4. Intake-Plans and Related Documents Submittal (Step 2)

Some agencies require a pre-scheduled meeting to accept and take plans and applications. To help applicants with the submittal process, many agencies have a checklist for applicants to include with submittal to ensure a complete submission. Therefore, before submittal, one should ensure that the engineer, the designer, or the project manager reviews and checks all the items on the application. If an item doesn't relate to the project, mark not applicable (N/A) with a brief explanation, but do not leave it blank! See Appendix A for a sample submittal application from the Hillsborough County, Florida.

Here are examples of some of the documents to be included for submission:

1) Application form
It usually includes owner(s) names and contact person names, property information, the scope of work, valuation, etc.
2) Owner's affidavit to authorize the applicant to represent the owner
3) Construction plans
4) Survey of the property (if required)
5) Geotechnical report (if required)
6) Technical information report (TIR) or stormwater management report

10.5. Review Process (Step 3)

After receiving the complete applications and associated documents, the permit center forwards all to the reviewers for review. For larger and more complex projects, and based on the size and complexity of the AHJ, one to a dozen divisions or departments may be involved in the review process. Different agencies also have processes and procedures to share the submitted documents with the reviewers. Some agencies may work like a clock and have a very efficient process in place, and others do not so. It also relates to the number of staff/reviewers for the agencies, their experience, the number of years with the same agency, etc.

Reviewers have days to weeks to complete their review and communicate their findings with the permit center. After receiving all the reviewers' comments, the permit center forwards them to the applicant for implementation and resubmittal.

Some agencies are very restricting and establish deadlines for how long the applicant can take before resubmittal. Other agencies are not. Regardless, it is essential to remember to respond to all questions, concerns, and comments in a professional and timely fashion. If in doubt, contact the reviewer to ensure that the questions are understood before assuming and responding accordingly. Another vital factor is that some comments or concerns may conflict with projects involving multiple agencies. The project manager's responsibility is to coordinate and collaborate with the reviewing agency to resolve any conflicting issue to avoid any approval delay and, worse, carry the unresolved issue to the field and cause construction delays.

10.6. Approval and Permit (s) Issuance (Step 4)

Step 3 above may take a few rounds of resubmittal to meet all AHJ codes, standards, and requirements to the reviewers' satisfaction. After receiving everyone's approval, the permit center gets ready to issue a permit. Before doing that, the permit center calculates additional permit fees based on the AHJ fee schedule. The following provides a list of the more common fees that must be paid before picking up the permit.

1) Additional review fee
2) Inspections fees (this may include inspections during construction and the warranty period, which could take from one to three years)
3) Traffic impact fee (this is typically collected upon issuance of the building permit)
4) Other impact fees may include school, stormwater, water, wastewater, etc.

In addition, the following documents may be required based on the type of permit to be issued.

1) The contractor licenses
2) Liability insurance
3) Posting bond or cash set aside as a security insurance (See Section 10.13. Financial Guarantees)

10.7. Construction Activities and Inspections (Step 5)

This step usually starts with a "job start" phone call or another mode of communication by the contractor to the AHJ, informing the inspector of the day and time the contractor plans to start the job and requesting a potential preconstruction meeting. Depending on the size and complexity of the project, the site inspector may ask for all the erosion and sediment control devices to be in place before the preconstruction meeting. The TESC is a critical element in every project but depending on the topography, soil condition, etc., it may be more dominant in some localities than others.

A preconstruction meeting provides an opportunity for all stakeholders (the contractor, owner, utility companies, inspector, and reviewers) to discuss any potential challenges to the project. Again, the bigger and more complex the project, the more significant the number of stakeholders involved. This meeting also allows the inspector to go over items that the inspector expects, and are usually overlooked by the contractor, that are not so obvious from the construction plans or things the inspectors emphasize are important to the AHJ. As an example, although access to residents must always be maintained, there could be residents with unique challenges that the inspector likes to bring to the contractor's attention.

After the preconstruction meetings and reviewing plans to ensure all stakeholders are on the same page before the construction start, in addition to inspections at certain stages of the construction progress, the inspector may stop by to inspect at their discretion and availability.

10.8. Correction Notices

Although correction notices happen on almost every job, the contractor must take them very seriously and address all corrections promptly and professionally to maintain a good relationship with the inspector and the AHJ. A more significant number of correction notices and delays to address issues creates a hazardous and distrustful environment between the AHJ and the contractor. That is never a good thing, especially for the contractor and the project's success. If in doubt, at least schedule a meeting with the inspector on-site to show that, as a contractor, you take correction notices seriously and make every effort to avoid them.

10.9. Stop-Work-Order (SWO)

Stop work orders mean all construction activities permitted by the AHJ must come to a halt. SWO can be very disruptive and expensive to the project's success. It usually happens after the issuance of numerous correction notices, and not getting the issues addressed by the contractor or their sub-contractors promptly and professionally. It also can occur if a very critical issue, such as a safety concern, is a significant concern to the inspector. A correction notice may not speedily get the contractor's attention to correct the deficiencies. After all, everything else should take a back seat when it comes to safety.

At this stage, it behooves all stakeholders to meet and discuss issues, remove the stop work order, and ensure the project can proceed smoothly without any additional hiccups. After addressing all necessary corrections and removing the stop work order, construction activities can start and hopefully move forward to completion. It is vital the parties not only focus on the issue at hand but also concentrate on upcoming challenges and try to come up with a systematic approach (i.e., step-by-step processes, better communication, and contact information) to resolve issues before it gets out of hand.

10.10. Final Site Inspections (Step 6)

At this stage, the contractor calls for the final inspection if the contractor and all subs completed their work per the AHJ-approved construction plans, permit conditions of approval, and to the satisfaction of the site inspector. They may schedule a final site walk to reinspect the entire site. During the walk-through, the inspector points to all deficiencies (if any) and the contractor. The contractor either provides reasons for shortages or accepts to complete those promptly and professionally. It is a good practice for the inspector to follow with an e-mail or letter and include a punch list of all items discussed and required to be addressed by the contractor. If the contractor doesn't receive a punch list, they may provide a list as they understand them to the inspector and ask the inspector to confirm for proper documentation. After completion of the punch list and another site inspection, if all are completed, the site receives the final from the inspector, and the contractor obtains a building certificate of occupancy, assuming there are no outstanding items related to the building to be occupied.

Please note the above-stated process includes both the construction activities on private property and the public ROW. Also, even though this step is final, we discuss the other required steps below.

10.11. As Built or Record Drawing

Many jurisdictions require an as built or record drawing to be submitted by the EOR to certify that the site was constructed substantially following the approved and permitted construction plans. The intent is to ensure that someone in charge of the design and construction plans preparation, preferably the EOR, oversees the construction and if the EOR and the AHJ approve any revisions. Finally, the EOR certifies that the constructed features will function per the design to the best of their knowledge.

Appendix A provides an as-built certification and request for conversion to the operation phase form. Submittal of this form is required by the SWFWMD after successful construction of any project approved by a stormwater management permit. Not all agencies have such a specific form. It is included here since much of the information requested on this form may be beneficial to other agencies to obtain.

The form mentioned above applies only to the stormwater management system. The EOR must submit this form on all construction projects with a stormwater management system that obtained a SWFWMD construction permit. The Department of Environmental Protection agencies has similar documents for constructed water and wastewater system construction and approval. Other agencies permitted the same project may accept these forms or have a separate document that must be completed and submitted to obtain their buy-out. See Appendix A for a sample of as-built form from the SWFWMD, Florida.

10.12. Certificate of Occupancy

A certificate of occupancy is issued to the owner after all required building and site improvements are completed, and the various inspection requirements are passed. The building and building-related inspectors inspect the entire building, including the architectural features, structure, and MEP (mechanical, electrical, and plumbing). The Fire department inspector inspects all fire-related elements such as sprinkler systems, fire standpipe, etc. This book is not about the building or building-related components. We mention them here because the certificate of occupancy cannot be issued without passing all those inspections. However, suppose all those building-related elements were constructed per the approved plans and passed the various inspections. In that case, the certificate of occupancy will only be issued if the final on-site engineering is also passed, and all outstanding issues are resolved.

The purpose here is to ensure that all site elements, such as access to the site, parking area, stormwater management system, and other utilities, are constructed per the approved construction plans and passed inspections. In other words, the site must be safe and secure before the owner can occupy the building.

10.13. Financial Guarantees

For all large residential sites other than single-family residences and commercial or industrial properties, the permitting agencies require that the owner provide some security device to guarantee the successful and timely completion of the construction site. They must post one of the following forms of financial guarantees at the time of permit issuance.

The purpose of the financial guarantee is to ensure timely and proper completion of improvements; ensure compliance with the AHJ code and standards and regulations. The owner is required to provide a secured warranty of materials and the quality of improvements. "Performance guarantee," "maintenance guarantee," and "defect guarantee" are considered subcategories of financial guarantee.

1) Performance Guarantee

At the time of permit issuance and prior to the AHJ's approval and construction authorization, the applicant shall provide the AHJ with a financial guarantee in the amount and in a form approved by the AHJ to secure compliance with all terms and conditions of the approval.

The financial security aims to ensure that the applicant will construct all proposed construction improvements (i.e., roadway, stormwater management system, and utilities) in full compliance with the approved construction plans and profiles. In addition, the construction meets or exceeds the requirements of the applicable AHJ ordinances, standards, and specifications to the satisfaction of the AHJ.

The financial guarantee shall remain in force and effect until written release by the AHJ. The owner further guarantees that if the subdivision, the short plat, or other required legal instruments are not recorded, the site does not receive a final construction approval, or the building does not receive a certificate of occupancy. The applicant agrees to restore the public ROW to its original condition or better. In addition, the applicant agrees to stabilize and secure the site or comply with the requirements of any permit or approval, including, but not limited to, corrective actions necessary to provide drainage consistent with approved plans and conditions and to protect the public health, safety, and welfare, including effects on water quality.

Most agencies provide a fillable spreadsheet to the applicant to compute the total construction cost of the facilities the agencies have interested in, called Bond Quantity Worksheet (BQW). Some agencies also accept an actual contractor's construction bid in place of the BQW. Regardless, the AHJ reviews the estimated construction cost to approve. After approval, the applicant must provide a performance financial guarantee based on the AHJ percentage before picking up associated permits to start construction. The performance guarantee amount could be anywhere from hundred percent to 150 percent of the actual construction cost of the onsite and offsite improvements. See Appendix A for a sample of BQW from the King County, Washington.

2) Operation and Maintenance Warrantee

Many projects require additional maintenance and operation inspections about one to three years after the successful completion. The purpose is to ensure that the completed project operates as intended. There were no materials, workmanship, or design deficiencies after the project went into operation.

Upon completing the required public improvements to the satisfaction of the AHJ, the applicant shall request the release of the above-said performance guarantee. The owner provides financial warranties and agreements in the amounts and forms acceptable to the AHJ. to secure the successful operation and maintenance of the improvements (i.e., roads, landscaping, stormwater management system, utilities) for a period accepted by the AHJ after construction approval.

The operation and maintenance guarantee amount could be anywhere from ten percent to fifty percent of the actual construction cost mentioned above. This computation is much easier than the previous one. Depending on the AHJ requirements, a percentage of the true performance guarantee is set aside before obtaining a certificate of occupancy or a final site construction project.

The following provides a brief description of different types of financial guarantees in no specific order:

1) Cash Deposit or Cash Set Aside

A cash financial guarantee works better for those companies with lots of cash. They don't have to go through the paper works and approval from the financial institutions or the bond company. They set aside money for the required amount of deposit. The cash deposit can't be terminated or canceled by the applicant until the written release of their obligation by the AHJ. Therefore, it may work better for the AHJ.

2) Irrevocable Standby Letter of Credit

An irrevocable standby letter of credit is a letter of credit issued by a financial institution and shall have the effective start date, or immediately, and the expiration date. It must remain in full force and effect until it expires per the AHJ permit terms and conditions. The expiration date must be after the completion of the construction project to the satisfaction of the AHJ.

3) Set Aside a Construction Loan Letter of Credit:

Setting aside a construction loan letter of credit is a less common form of construction loan credit that the loan company provides to the AHJ.

4) Surety Bond:

A surety bond is probably the most common form of performance and maintenance financial security the owner provides. The owner pays a monthly or annual fee to obtain such financial security for as long as the project is under construction. Depending on the project and the owner's credit situation, it may take a few weeks to obtain.

This form of financial security sometimes creates additional work or hardship for the AHJ. If the owner defaults in paying the fee, the bond company cancels the bond, which could happen at any time during the construction.

Other forms of financial security that may be acceptable to the AHJ.

10.14. Other Approvals or Permits

Other reviews performed by the AHJ may not require permits but are typically triggered or part of another type of permit. The following sections address those areas that require approval from an AHJ but may not be a stand-alone permit:

1) Noise Variance or Approval

Almost all municipalities have some code to address noise levels, particularly at night. What counts as nights is generally defined by a municipal code or standard specifications. Essentially, any construction and other related activities that generate noise above a certain decibel (i.e., normal conversation 55-65 dB, jackhammer and power saw 110 dB) allowed by the municipalities require some approval or permission from the AHJ before starting the work. For example, an agency may define working hours that impact the residential zones as the following restricted hours:

Monday through Friday: 7 a.m. to 7 p.m.
Saturday: 9 a.m. to 6 p.m.
Sunday or legal holidays: Prohibited

The noise variance may allow the working hours to be extended outside the 7 a.m. to 7 p.m. on Mondays through Fridays and outside the 9 a.m. to 6 p.m. on Saturdays. Therefore, construction activities outside those hours require an approved noise variance or approval. The approval of the noise variance highly depends on the location of the project and its proximity to the residential neighborhood or commercial hotels, and the noise level generated by the construction activities.

Getting a noise permit is more straightforward than dealing with a stop work order for working outside the regular hours. Generally, this permit only requires an application with the applicant's contact information and the scope of the construction activities form for submittal because it will already be associated with another type of permit with plans. It is important to remember that the work will be stopped, regardless of what permission is in hand if the noise level generated by the construction activities disturbs the public's peace. It behooves the applicant to do anything in their power to ensure public safety and comfort.

Regardless of a noise variance approval, if the AHJ receives complaints regarding the noise level, the noise variance will be voided and denied.

2) Off-Hour Work or Overtime Approval

Typically, an off-hour request accompanies a noise variance for night work, but it may be for a few hours at the end of the day and not require a noise variance. For example, most construction companies start early and stop for the day around three or 3:30 p.m. to beat the heavy traffic. If they are trying to catch up and work till 5:00 p.m., there will be no need for noise variance, but they may have to get an off-hour work approval if there is a need for an inspector to inspect the construction activities.

If the job requires inspection by the AHJ staff, the agency reviews the requested off-hour request to see if they have the right inspector to inspect the work during the requested hours. The agency may deny the permit if no inspector can inspect the work activity. Also, any work inspected during off-hour work qualifies for overtime, especially if the worker is under the union. The applicant agrees to pay the overtime based on the accepted hourly rate to include the overtime.

3) Permit Extension or Cancellation

Usually, this simple form asks to change an application's status or expiration date. As the title implies, there are times that the applicant needs to cancel the application and sometimes extend the permit expiration date to finish the job. It is vital to remember that to extend the expiration date, the applicant must describe the situation causing the delay. Also, there may be limits on the number and duration of an extension. The best time to submit this form is before the permit expires. Preferably two to four weeks in advance, allowing the AHJ to review and see how to handle the request. Sometimes, an applicant may be qualified to get back a partial permit fee to cancel the permit, depending on the AHJ and the status of the permit.

CONGRATULATIONS!

If you have any comments or questions, please complete the form at the end of this book and e-mail it to me.

BEST WISHES,
Ali

ACRONYMS AND DEFINITIONS OF TERMS

AASHTO: American Association of State Highway and Transportation
ADU: accessory dueling unit aka mother-in-law unit
AHJ: agency having jurisdiction (i.e., city, county, state)
AKA: also known as
AOI: area of interest
ASTM: *American Standards Testing Manual*

BMP: best management practice
BSBL: building setback line

CESCL: certified erosion and sediment control lead
CN: curve number
CPESC: certified professional in erosion and sediment control
CPTED: crime prevention through environmental design

DBH: diameter breast height
DOE: Department of Ecology
DOH: Department of Health

EPA: Department of Environmental Protection Agency

FAR: floor to area ratio
FDOT: Florida Department of Transportation
FEMA: Federal Emergency Management Agency
FHA: Federal Highway Administration

GIS: geographic information system
GWE: groundwater elevation

HOA: Homeowners Association
HP: Hewlett Packard
HSG: hydrologic soil group

ITE: Institute of Transportation Engineers

KCSWDM: *King County Surface Water Design Manual*

LID: low impact development

MOT: maintenance of traffic
MUTCD:

NGVD: National Geodetic Vertical Datum Officials
NOI: notice of intent
NPDES: National Pollutant Discharge Elimination System

OSHA: Department of Labor Occupational Safety and Health Administration

PODS: Potable on Demand Storage
PUD: planned unit development
PVC: polyvinyl chloride

ROW: right-of-way

SBL: setback lines
SCS: soil conservation service
SE, LLC: Shasti Enterprises, LLC
SFR: single family residential
SHWE: seasonal high-water elevation
SHWT: seasonal high-water table
SWFWMD: Southwest Florida Water Management District
SWMS: stormwater management system
SWPPPs: stormwater management

Tc: time of concentration
TCP: traffic control plan
TESC: temporary erosion and sediment control
TIA: traffic impact analysis
TIF: transportation impact fee
TIR: technical information report
TSS: temporary sediment sump
Tt: travel time

USDA: United States Department of Agriculture
USGS: United States Geological Survey

WSS: web soil survey
WSDOT: Washington State Department of Transportation
WWHM: Western Washington Hydrology Model
WWM: welded wire mesh

AEC, Inc.: Access Engineering and Consulting, Inc. established by M. Ali Shasti Nazem, P.E., MSCE, a site engineering, structural and environmental engineering operated in Plant City, Hillsborough County, and Zephyrhills, Pasco County, Florida from 1999 to 2011.

Applicant: A property owner or representative or any person or entity designated or named in writing by the property or easement owner to be the applicant in an application for a development proposal, approval, or permit.

Arterial road: A road or street primarily for through traffic. The term generally includes roads or streets considered collectors. It does not include local access roads, which are typically limited to providing access to abutting properties.

As built: Engineering plans which have been revised to reflect all changes to the projects that occurred during construction.

Assignment of funds: An assignment of funds is a process used by a financial institution to redirect funds from a line of credit to a third party. It must be approved by the financial institution that granted the line of credit following a request fulfilling any obligations by the AHJ.

Baffle: A device, usually a flow-directing or impeding panel, used to deflect, check, or regulate flow.

Best management practice (BMP): Any schedule of activities, prohibition of practices, maintenance procedure, or structural and managerial approach meeting approved construction plans, when used singly or in combination, prevents or reduces the release of pollutants and other adverse impacts to surface water, stormwater, and groundwater.

Bioretention: A flow control best management practice consisting of a shallow landscaped depression designed to store and promote infiltration of stormwater runoff temporarily. The designer must follow specific design standards, such as soil mix, plant species, storage volume, and feasibility criteria.

Bioswale or grassed swale: A long, gently sloped, vegetated ditch designed to remove pollutants from stormwater. The grass is the most common vegetation. For saturated soil, wetland vegetation is also used.

Building setback line: A line measured parallel to a property, easement, drainage facility, or buffer boundary that delineates the area (defined by the distance of separation) where buildings or other obstructions are prohibited (including decks, patios, outbuildings, or overhangs beyond eighteen inches). Wooden or chain-link fences and landscaping are allowable within a building setback line.

Catch basin insert: A device installed underneath a catch basin inlet that uses gravity, filtration, or various sorbent materials to remove pollutants from stormwater. When used with sorbent material, catch basin inserts are primarily for oil removal.

Certified erosion and sediment control lead (CESCL): An individual with current certification through an approved erosion and sediment control training program. The program must meet the minimum training standards established by the Department of Ecology (DOE) in Washington State and the Oregon Department of Environmental Quality (DEQ) in Oregon State. A CESCL is knowledgeable in erosion and sediment control principles and practices. The CESCL must have the skills to assess *site* conditions and construction activities that could impact the quality of stormwater and the effectiveness of erosion and sediment control measures used to control the quality of stormwater discharges. Certification is obtained through an approved environmental department erosion and sediment control course.

Cisterns: Containers of various sizes, below or above ground structures to store rainwater runoff delivered through building downspouts.

Civil engineer: A person licensed by the Board of Professional Engineers as a professional engineer in civil engineering in the state they are practicing.

Clear or sight visibility triangle: The sight visibility triangle, or sight triangle, consists of three sides formed by two intersecting access ways and a line connecting the two. The first side is drawn parallel to the centerline of the intersected street, and the second is drawn parallel to the centerline of the other street. The distances are a function of the posted speeds. Typically, the AHJ limits the heights of the structures, vegetation, and other improvements on corner properties immediately adjacent to intersections. The intent is to unblock the view of the drivers turning.

Closed drainage basin: A watershed in which the stormwater runoff does not have a surface outfall up to and including the hundred-year flood level.

Contractor: The person, partnership, firm, or corporation contracting to perform construction work under the contract documents. The term shall also include the Contractor's agents, employees and subcontractors.

Control device: The element of a discharge structure which allows the gradual release of water under controlled conditions. This is sometimes referred to as the bleed-down mechanism or "bleeder". Examples include orifices, notches, weirs, and effluent filtration systems.

Control elevation: The lowest elevation at which water can be released through the control device. This is sometimes referred to as the invert elevation.

Control structure "KAA Discharge Structure": A structural device, usually of concrete, metal, etc., through which water is discharged from a project to the receiving water.

Cul-de-sac: A short road or street having one end open to traffic and the other temporarily or permanently terminated by a vehicle turnaround at or near the terminus. It typically serves more lots vs. hammerhead turnaround which is for fewer lots.

Curve number (CN): The runoff property of a specific soil and the ground cover. A high CN value such as 98 is assigned to impervious surfaces, such as pavement and causes the most stormwater runoff with no infiltration. A lower value, such as fifty-eight for some wooded regions, corresponds to an increased ability of the soil to retain and infiltrate rainfall and produce much less runoff.

Detention: The release of surface and stormwater runoff from the site at a slower rate by collecting, storing, and delaying the discharge at a lower rate.

Detention volume: The volume of open surface storage behind the discharge structure measure between the overflow elevation and control elevation.

Developer: See applicant.

Deviation: See variance.

Drainage: The collection, conveyance, containment, or discharge, or any combination thereof, of stormwater runoff or surface water.

Drainage area: An area draining to the point of interest.

Drainage basin: A subdivision of a watershed.

Drainage easement: A drainage easement is a legal encumbrance placed against a property's title to reserve specified privileges for the users and beneficiaries of the drainage facilities contained within the easement.

Drainage facility: A drainage facility is a constructed or engineered feature that collects, conveys, stores, treats, or otherwise manages stormwater runoff or surface water. Drainage facility includes, but is not limited to, constructed, or engineered ponds, lakes, wetlands, water quality, flow control BMP, and any other structure and appurtenance that provides drainage.

Drywells: Gravel or stone-filled pits designed to catch and filter stormwater runoff from the roof downspouts or paved areas and infiltrate the ground.

Easement: An easement grants the right to use the real property of another for a specific purpose.

Elevation: The height in feet above the mean sea level according to National Geodetic Vertical Datum (NGVD).

Erosion: The detachment and transport of soil or rock fragments by water, wind, ice, etc.

Erosion and sediment control (ESC): The temporary or permanent measures taken to reduce erosion and control siltation and sedimentation to ensure that sediment-laden water does not leave the site or enter wetlands or aquatic areas.

Exposed: Not fully covered and subject to direct or blown-in precipitation and/or natural or blown-in runoff.

Filter strip: A dense vegetation area with gentle slopes which treats stormwater runoff from adjacent areas before entering a receiving structure or water body.

Financial guarantee: A form of financial security posted to do one or more of the following: ensure timely and proper completion of improvements; ensure compliance with the AHJ standards and requirements; or provide a secured warranty of materials and quality of design and construction improvements. Financial guarantees include assignments of funds, cash deposits, surety bonds, or other forms of financial security acceptable to the AHJ.

Flow control BMP: A small-scale drainage facility or feature part of a development site strategy. The Flow Control BMP uses processes such as infiltration, dispersion, storage, evaporation, transpiration, forest retention, and reduced impervious surface footprint to mimic pre-developed hydrology and minimize stormwater runoff.

Fully covered: Not exposed and covered sufficiently to shield from rainfall and stormwater run-off. At a minimum, full coverage requires a roof with enough overhang in conjunction with walls of sufficient height to prevent rainfall blow-in; and the walls must extend into the ground or to a berm or footing to prevent runoff from being blown in or from running onto the covered area.

Hammerhead turnaround: A short road or street having one end open to traffic and the other temporarily or permanently terminated by a vehicle turnaround at or near the terminus. It typically serves fewer lots vs. cul-de-sac, which is for more lots.

High-use site: A commercial or industrial site that typically generates or is subject to runoff containing high concentrations of oil due to high traffic turnover, on-site vehicle or heavy or stationary equipment use, or the frequent transfer of liquid petroleum.

Historic discharge: The peak rate and/or amount of runoff which leaves a parcel of land by gravity from an undisturbed/existing site.

Infiltration trenches: Trenches filled with porous media such as coarse or medium sand or aggregate that collect stormwater runoff from the roof or pavement and infiltrate it into the ground. They are a good option where the depth to the maximum wet-season water table or hardpan is between three and six feet.

Impervious surface: All hard or compacted surfaces like roofs, pavement, gravel, or any compacted surfaces that allow minimal or no infiltration to the ground.

Inlet pollution removal devices: Small stormwater treatment systems are installed below grade at the edge of paved areas to filter or trap pollutants in runoff before it enters the storm drain.

Land disturbing activity: Any activity that changes the existing soil cover, vegetative and non-vegetative, or the existing soil topography. Land disturbing activities include but are not limited to demolition, construction, clearing, grading, filling, excavation, and compaction. Tilling as part of agricultural practices, landscape maintenance, or gardening is not considered land disturbing activity.

Landlocked: In real estate, this refers to a piece of property that is surrounded by other properties and has no direct access via public thoroughfare (street, easement, etc.).

Legal access: Where each lot in a subdivision has access to a public (city, county, state, or federal) street or road, or the subdivider has obtained adequate and appropriate easements across all necessary properties from a public road to each lot in the subdivision.

LID best management practices: Distributed stormwater management practices that are integrated into a project design that emphasizes pre-disturbance hydrologic processes of infiltration, filtration, storage, evaporation, and transpiration. LID BMPs are flow control BMPs and include, but are not limited to infiltration, and limited infiltration systems, roof downspout controls, dispersion, permeable pavements, and bioretention.

Local government agency/local agency: Any government agency such as a town, village, city, or county.

Low impact development (LID): A stormwater and land use management strategy to mimic pre-disturbance hydrologic processes of infiltration, filtration, storage, evaporation, and transpiration by emphasizing conservation, on-site natural features, site planning, and distributed stormwater management practices integrated into a project design.

Mitigation: An action or series of actions to offset the adverse impacts that would otherwise cause a regulated activity to fail to meet the AHJ requirements. Mitigation usually consists of restoration, enhancement, creation, preservation, or a combination thereof.

Open drainage basin: Open drainage basins are all watersheds not meeting the definition of closed drainage basin defined above.

Overflow elevation: The design elevation of a discharge structure at or below which water is contained behind the structure, except for that which leaks or bleeds out, through a control device down to the control elevation.

Permeable pavement: Pervious concrete, porous asphalt, permeable pavers, or other forms of pervious or porous paving material intended to allow the passage of water through the pavement section. It often includes an aggregate base that provides structural support and acts as a stormwater reservoir.

Pervious (porous) surface: All those soft or non-compacted surfaces like forests, meadows, pastures, or lawns that allow the stormwater runoff to infiltrate the ground.

PES, LLC: Professional Engineering Services, LLC established by M. Ali Shasti-Nazem, P.E., MSCE, a land planning and civil engineering firm operated in Kent, King County, Washington from 2012 to the present.

Rain garden: A shallow, landscaped depression with compost-amended native soils and adapted plants. The depression is designed to detain and temporarily store stormwater runoff from adjacent areas and to allow stormwater to pass through the amended soil profile.

Rain harvesting: Rainwater harvesting is collecting and storing rainwater for various uses, including but not limited to irrigation, drinking and domestic consumption, and stormwater runoff reduction.

Recovery time: It is the time that the stormwater management storage facility takes to recover and be ready to accept additional stormwater runoff for the designed storm event.

Restoration: Converting back to a historic or pre-development condition.

Retaining wall: A retaining wall is an engineered or non-engineered structure that retains (holds back) materials such as earth and prevents it from sliding or eroding. Typically, a retaining wall taller than four feet requires a building permit.

Retention: The prevention of direct discharge of storm runoff into receiving waters; included as examples are systems which discharge through percolation, exfiltration, and evaporation processes.

Seasonal high-water level: The elevation to which the ground or surface water can be expected to rise due to a normal wet season.

Sediment: Fragmented material that originates from weathering and erosion of rocks or unconsolidated deposits and is transported by, suspended in, or deposited by water.

Setback line (SBL): A line measured parallel to a property, easement, drainage facility, or buffer boundary that delineates the area (defined by the distance of separation) where buildings or other obstructions are prohibited (including decks, patios, outbuildings). Typically, wooden, or chain-link fences and landscaping are allowed within a building setback line.

Sight visibility triangle: See clear visibility triangle.

Simplified drainage design: A simplified form of site improvement and erosion and sediment control plans (without technical information or drainage report) prepared by a non-engineer from a set of pre-engineered design details.

Soil amendment: Minerals and organic material added to the soil to increase its capacity for absorbing moisture and sustaining vegetation.

Stormwater management system: A system designed and constructed or implemented to control discharges which are necessitated by rainfall events, incorporating methods to collect, convey, store, absorb, inhibit, treat, use, or reuse water to prevent or reduce flooding, over drainage, environmental degradation, and water pollution or otherwise affect the quantity and quality of discharges from the system.

Sub basin: A subdivision of a basin.

Surface water management system facilities: The surface water management system facilities shall include, but are not limited to, all inlets, ditches, swales, culverts, water control structures, retention and detention areas, ponds, lakes, floodplain compensation areas, wetlands, and any associated buffer areas, and wetland mitigation areas.

Track-out: The mud, soil, or other organic debris from a construction site, farm, or other areas where the ground has been disturbed.

Tract: A portion of land with defined boundaries created by a deed.

Traffic control supervisor: A qualified person with some level of American Traffic Safety Services Association (ATSSA) training and credential. Various departments of transportation offer the same type of training and certifications.

Traffic impact analysis (TIA): Traffic study and report performed by a qualified traffic engineer or traffic engineering company using ITE.

Tree box filters: Curbside containers placed below grade, covered with a grate, filled with filter media, and planted with a tree in the center.

Variance: A variance, waiver, or deviation is a form of approval by the AHJ to allow certain deviations from the agency's code, standard, or requirements. The requirements for requesting such deviations may significantly differ between the AHJ.

Vault: A stormwater detention or retention vault is an underground pre-cast or cast-in-place structure designed to store stormwater runoff on a developed site, typically in an urban setting, due to limited land availability.

Vegetated roofs (aka green roofs): Consist of a previous growing medium, plants, and a moisture barrier. They reduce the stormwater runoff peak and volume due to their soil storage capabilities and increased evapotranspiration rate.

Waiver: See variance.

Water management areas: Areas to be utilized for the conveyance or storage of surface water, mitigation, or perpetual operation and maintenance purposes.

Watershed: Contains all the land areas that contribute to water flow into a receiving body of water.

Weir: A control or discharge structure that regulates water flow from storage facilities such as detention, retention pond, or vault.

Welded wire mesh: Welded wire mesh or fabric is a prefabricated joined grid consisting of parallel longitudinal wires with accurate spacing welded to cross wires at the required spacing. One of the uses is to reinforce concrete driveways.

Well-drained: Well-drained soil is any soil that allows water to drain through it quickly, such as sand.

Wet detention system: A water quality treatment system that utilizes a design water pool in association with water-tolerant vegetation to remove pollutants through settling, adsorption by soils and nutrient uptake by the vegetation. The bottom elevation of the pond must be at least one foot below the control elevation.

Wheel wash: A wheel wash system is used at the exits of the construction sites, quarries, waste management, and other industrial areas to prevent dirt and debris from ending on public roads and highways. The wheel wash washes the wheels before exiting the site.

REFERENCES

"ASTM. American Standard Testing Manual." 2018. All Acronyms. https://www.allacronyms.com/ASTM/American_Standard_Testing_Manual.

Barratt, Claire and Whitelaw, Ian. *The Spotters Guide to Urban Engineering: Infrastructure and Technology in the Modern Landscape.* Cape Town: Zebra Press, 2011.

"Building Construction Services." Pasco County, FL. 2020. https://www.pascocountyfl.net/283

Colley, Barbara. *Practical Manual of Land Development.* New York: McGraw-Hill, 2005. "Department of Transportation (DOT)." Washington State DOT. 2018. https://www.washington-agencies.com/department-of-transportation

"Departments." City of Lakeland, FL. 2018. https://www.lakelandgov.net/departments "Departments." Hillsborough County, FL. 2020. https://www.hillsboroughcounty.org/en/government/departments

"Departments, agencies, & offices." King County, WA. 2018. https://kingcounty.gov/depts "Development Services." City of Bellevue, WA. 2021. https://bellevuewa.gov/city-government/departments/development

"Development Services Center." Redmond, WA. 2019. https://www.redmond.gov/898 "Information Site." Manual of Uniform Traffic Control Device (MUTCD). 2020. https://mutcd.info

"Flood Maps." FEMA.gov. 2018. https://www.fema.gov/flood-map

Johnson, David. *Residential Land Development Practices A Textbook on Developing Land into Finished Lots, Fourth Edition.* Reston: ASCE Press/American Society of Civil Engineers, 2008.

Kone, Daisy. *Land Development.* USA: BuilderBooks, 2006.

McCormac, Jack. *Surveying Fundamentals.* Englewood Cliffs, N.J.: Prentice Hall, 1976. Merritt, Frederick and Ambrose, James. *Building Engineering and Systems Design.* New York: Van Nostrand Reinhold, 1990.

Merritt, Frederick and Ricketts, Jonathan. *Standard Handbook of Civil Engineers.* New York: McGraw-Hill, 1986.

"Occupational Safety and Health Administration (OSHA)." 2020. United States Department of Labor. https://www.osha.gov

Papacostas, C. S. *Fundamentals of Transportation Engineering.* USA: Prentice-Hall, 1990. "Permit Center." City of Kent, WA. 2020. https://www.kentwa.gov/pay-and-apply/apply-for-a-permit

"Permit Information." City of SeaTac, WA. 2021. https://www.seatacwa.gov/how-do-i/ apply-for/a-permit

"Regulations & Permits." Washington State Department of Ecology. 2019. https://ecology.wa.gov/Regulations-Permits

"Soil Survey Manual." Natural Resources Conservation Service. 2018. https://nrcs.rancher.usda.gov/resources/guides-and-instructions/soil-survey-manual

South Florida Water Management District (SWFWMD). 2020. https://www.sfwmd.gov "The Home of Transportation Professionals." AASHTO. https://www.transportation.org

"Traffic Engineering." Institute of Transportation Engineers. 2020. https://www.ite.org/ technical-resources/topics/traffic-engineering

"U.S. Geological Survey." USA Gov. 2020. https://www.usa.gov/federal-agencies/u-s-geological-survey

"Web Soil Survey." Natural Resources Conservation Service. 2020. https://data.nal.usda.gov/dataset/natural-resources-conservation-service-web-soil-survey

"Urban Hydrology for Small Watersheds." TR-55 - National Stormwater Trust. 2016. https://nationalstormwater.com/urban-hydrology-for-small-watersheds-tr-55

"10 States Standards - Recommended Standards for Wastewater Facilities." Atlantic Canada Water & Wastewater Association. Albany: Health Research Inc. https://www.acwwa.ca/.../design-guidelines/189-10-states-standards-wastewater-2004/file

READER COMMENTS FORM

Thank you for choosing *SITE DESIGN and ENGINEERING.*

Your professional feedback and assistance in completing this form are greatly appreciated. Your honest feedback will help us to improve the next edition. Thank you, and best wishes.

Reader name: ____

Title (optional): ____

E-mail address: ____

Phone # (optional): ____

Your comments: Please refer to the section or page number

____
____
____
____
____
____
____
____
____
____
____
____
____
____
____
____
____
____
____
____
____

Please e-mail this form to AShasti13@Gmail.com

9 798987 093764